To Piedmont College

Hugh Macaulay

**Environmental Use
and the Market**

# Environmental Use and the Market

**Hugh H. Macaulay**
**Bruce Yandle**
Clemson University

**Lexington Books**
D. C. Heath and Company
Lexington, Massachusetts
Toronto

**Library of Congress Cataloging in Publication Data**

Macaulay, Hugh Holleman, 1924–
    Environmental use and the market.
    Bibliography: p.
    Includes index.
    1. Externalities (Economics) 2. Environmental policy. 3. Pollu-
tion—Economic aspects. I. Yandle, Bruce, joint author. II. Title.
HB99.3.M2        330'.01        77–91
ISBN 0–669–01431–1

Second printing, October 1978

Published simultaneously in Canada

Printed in the United States of America

International Standard Book Number: 0–669–01431–1

Library of Congress Catalog Card Number: 77–91

# Contents

# List of Figures

# List of Tables

# Preface

During the sixties, several social causes became modern substitutes for the search for the Holy Grail and the recapture of the Holy Land. While two centuries were necessary for eight or more crusades to retake Jerusalem in the early part of this millenium, at least three major marches were mounted in the recent decade. People sought cures for racial prejudice, poverty, and environmental degradation. Of the three, the last offers the best prospect for a reasonable and widely accepted solution. Yet, like so many cases in which emotion rather than reason guides the way, the cure may consist of substituting one social ill for another. Our collective concern for purity appears to have led us to create more problems than we have solved.

People are worried about our environment, and properly so. It is the only one we have or are likely to have in the foreseeable future. But the environment is more than fresh air and sunshine. It also takes corn flakes, cotton frocks, and Curl Free, as well as camp fires, to provide humans with their most pleasurable surroundings. A broader view of the environment is necessary, and in our book, we attempt to provide that view. If society is going to seek some goal, it helps to choose a useful goal and to make it clear to the public why the goal is worth pursuing.

Much of the literature on the environment stresses the problem of pollution. Indeed, with sewage in the water, smog in the air, and sound everywhere, it would appear obvious that pollution is *the* problem. Yet, we argue that a preoccupation with pollution can lead us astray, and perhaps has done so. Society has become concerned with stopping pollution and compensating those damaged by it. However, if we view the problem not as a matter of pollution but as a question of how to use the environment, two odd results follow. First, the aim becomes not to stop pollution but to make sure we have some pollution; second, we need not compensate people who consider themselves hurt by pollution but ask them to pay for not being hurt more.

These appear to be odd, even ridiculous, conclusions. However, they agree with what we do voluntarily every day. For example, most of us consider work as an activity akin to pollution, which we would like reduced to zero. But it is better to have some work than no work. As for compensation, when someone bids higher for the land we use or the gardener we employ, we may argue that we should be compensated for the loss we suffer from higher prices we must pay for these items; but usually, if we still buy, we pay more. We do not receive even a card of sympathy.

By understanding the way the market works, we can understand better the way the environment can be made to work for the welfare of mankind. There are environmental problems deriving from externalities, property rights, and even price, the very foundations of markets. We must examine and understand each of these if we are to deal effectively with the environment. Once mastered, these concepts have broad applicability in solving problems. Wherever two or more people are gathered together and they differ on some matter, they can use these tools to determine an optimum solution. The tools may help people from the Far East who seek to find a balance in the forces of nature; and they may help college sophomores as they search for their oft-cited Happy Medium. "Should we play the National Anthem at public events? Should abortions be performed? How should cities be zoned?" All these questions can be analyzed more easily and accurately if we understand the application of these tools to environmental problems.

Two questions often arise in discussions of the environment: "Who should pay for environmental improvements?" and, "What about the long-run consequences of environmental pollution?" These are covered in our book, along with many questions that are not asked quite so often. The questions about the environment are genuine, and they are serious. We hope the answers will be clear and that they will contribute to man's ability to gain the greatest enjoyment from his environment.

Our efforts to apply the concepts of the market to environmental problems draw on the contributions of many colleagues and students who listened to and criticized our ideas; research organizations which provided funds for our investigations; the dean of our college who provided an atmosphere which fostered research; and, of course, our first mentors who focused our attention on the power of economic analysis and guided our graduate research. Accordingly, we express our deep appreciation to our associates in the Department of Economics at Clemson University; to the Office of Water Resources Research of the Department of the Interior and South Carolina's Water Resources Research Institute; to Dean Wallace D. Trevillian of Clemson's College of Industrial Management and Textile Science; and finally to our dissertation chairmen, Professor Carl S. Shoup and Professor Richard A. Bilas. While we gratefully acknowledge this help and inspiration we accept and share the responsibility for the final product.

# 1        Introduction

## The Environment as an Economic Problem

If we collated those subjects of greatest concern to the inhabitants of North America and the rest of the industrialized world, surely environmental quality, or "environmental pollution" as it is more likely to be called, would place in the top three to five. It would rank along with peace (or "war" as it is more likely to be designated) and economic prosperity (or "economic recession" as it is more likely to be described). As we shall see later, the way we describe a problem can help or mislead us when we try to solve it.

Nonetheless, this rise to prominence has been rather sudden. Thirty years ago, J. Frederic Dewhurst gave us a detailed description of *America's Needs and Resources* (1947), but only about one page out of almost 800 dealt with pollution, and that page was confined to water pollution in certain river basins (p. 447). Despite the absence of electrostatic filters, scrubbers, and catalytic converters, the need for cleaner air did not rate an entry in the extensive index. Further, Allen Kneese and Charles Schultze (1975) remind us that when the American Assembly discussed Goals for Americans in 1960, it listed fifteen such goals, but improved environmental quality was not one of them (p. 2). We are inclined to wonder how we fell so far so fast.

The rise in concern for environmental quality from that relatively low level to its present elevated status can be seen more precisely by referring to that semi-official chronicler of events at home and abroad—the *New York Times*. The annual *Index* of that paper records succinctly not only the march of time but also the rise and fall of current concerns and fads. Table 1–1 reveals some of the elements dealing with the environment.

It appears that our interest in water exceeds our interest in air. Now sound and solid waste are ascending while water is waning. Interest in strip mining is on the rise. The seventies seem to be the decade of varied environmental decadence.

The role of economics in dealing with environmental problems has been somewhat slow in developing,[1] but the rate has accelerated markedly in the last decade. We shall attempt to show that although problems relating to the environment should fit easily into existing economic theory, in effect new theories have been developed to deal with them. It appears that the earlier economic theories were sufficient to answer the

**Table 1–1**

**Pages in the New York Times Index Devoted to Stories Dealing with Environmental Pollution**

| | | | Pages Devoted to Stories Dealing with | | |
|---|---|---|---|---|---|
| Year | Water Pollution | Air Pollution | Noise | Waste Disposal on Land | Total Pages |
| 1954 | .33 | .67 | .30 | .30 | 1223 |
| 1956 | .25 | .50 | .50 | 1.00 | 1424 |
| 1958 | .50 | .50 | .25 | .44 | 1025 |
| 1960 | .50 | .67 | .25 | .16 | 1131 |
| 1962 | .75 | .50 | .33 | .11 | 1062 |
| 1964 | 1.25 | .67 | .33 | .11 | 1189 |
| 1966 | 2.00 | 2.00 | .67 | .35 | 1396 |
| 1968 | 2.33 | 2.00 | .33 | 3.33 | 1710 |
| 1970 | 17.50 | 12.80 | .33 | 3.80 | 2304 |
| 1972 | 12.40 | 8.67 | 1.00 | 4.00 | 2564 |
| 1974 | 8.25 | 7.33 | Subdivided 30 Headings | 5.00 | 2828 |

questions that arose, but the stress on special treatment of the environment has led us into paths of environmental unrighteousness.

Furthermore, while special theories designed to answer questions about environmental pollution have been developed, there has been a tendency to broaden their application to include more traditional problems. Land, food, gasoline, and labor are being treated as if they are subject to the new economic theories of pollution. It appears that treating these goods as if they, too, require special economic theories has led us into paths of economic unrighteousness.

The upshot is that we have discovered some "new" aspects of the environment. But what we have found is not really new. The conditions have been with us for centuries and have produced before the same problems that we now consider unique to our age. In the past, the problems associated with the environment were handled with classical economic theory, and there is no reason to doubt its potential effectiveness today. But because environmental use has been viewed as something new, a new theory has been invented to handle the new problems: and that theory has been extended to the old economic questions. Parts of the theories are not efficient in dealing with either. In a sense, it is new wine in old bottles. The chapters which follow form the justification for these statements.

## Some Definitions of the Environment

Outside of the brief sojourn in the Garden of Eden, humanity has been beset with environmental problems. People have sought clean water, fresh air, warm sunshine, and fertile soil, all rolled into one happy hunting ground. Such combinations, however, have been hard to find. Usually, when people located a place where they wanted to live, so did insects, lions, snakes, fungi, germs, and a general run of unwelcome guests, including great white sharks. Until about the nineteenth or twentieth century, people often tried to improve their environmental lot by moving on to a better site. As Kenneth Boulding (1966) describes it, man lived in a Cowboy Economy (p. 9). If he did not like his present surroundings or if they became polluted, he could pack up his gear and move on to greener pastures or prairies. Today, conditions are different. There are fewer unoccupied, uncluttered, and unpolluted places to which people can move, and possibly, wherever people choose to move, the polluted air they seek to leave will follow them around the globe. Boulding characterizes man today as living in a Spaceship Economy. There is nowhere to throw the garbage without dumping it on someone else.

As has been true throughout the ages, we tend to think that our problems are the most serious that history has known. Such observations usually tell us more about man's egotism and perception than they do about the problems under discussion. However, there is always the chance that although the problem may not be fatal to mankind, it is at least worth the effort spent to ameliorate it. The current concern over the environment is voiced not only by people in the streets who announce that the end is nigh, but also people in the halls of government, the halls of academe, and even the hiring halls. With so many saying so much, it helps to know what the problem is in order to deal with it.

The concept of the environment is a good place to begin. Despite the precision that we like our language to provide, many English words have more than one meaning. Environment appears to fall into that classification. The *Oxford English Dictionary* cites as a concrete meaning of environment, "That which environs; the objects or the region surrounding anything," and then elaborates, "*esp.*[ecially] the conditions under which any person or thing lives or is developed; the sum-total of influences which modify and determine the development of life or character."[2]

A comparison of the second and third editions of the Webster's Unabridged Dictionary provides a hint about the change taking place in our use of the term "environment." The second edition gives as one definition, "That which environs; the surrounding conditions, influences, or forces, which influence or modify," and follows this with several specialized uses that are specifically designated, such as biological environment and social environment. The third edition uses virtually the same definition and follows with two examples. The first stresses biological and natural forces, while the second stresses social and cultural forces.[3] However, it appears that environment can refer to either of these conditions without special designation. Previously, if reference was to a restricted meaning, it was so stated. Today special designations are not given, and it is possible that two people talking about the environment may be talking about two different things.

A good example of this confusion occurred at public hearings on proposed air quality regulations for the State of Hawaii, held in Honolulu on January 12, 1972 (attended by one of the authors). The hearings dealt with a proposal to ban the burning of sugarcane in the fields in order to reduce the amount of smoke in the air and so to improve the environment. However, opponents argued that this would also make it uneconomical to grow sugarcane in Hawaii because of the additional processing that would result. One of the workers from the sugarcane fields pointed out that he lived near the fields, and the air he breathed had much more smoke in it that did the air breathed by the residents of distant Honolulu. But he also noted that if the smoke were banned, his environment would not improve but would deteriorate, and he would be much worse off. He would be without a job, and though he might have cleaner air, he would have fewer worldly goods. The "sum-total of influences" on his life would give him a less desirable environment. He referred to the general meaning of "environment"; the residents of Honolulu were concerned only with the more limited meaning dealing with biological and natural forces.

Here lies part of the problem with environmental quality. When one person speaks of the environment, he may refer only to the air, or perhaps the water, or both. To him, an improved environment means cleaner air and water. Another person sees environment as everything that surrounds him. When promised a better environment, he perks up in happy anticipation. If he gets cleaner air and water, but less heat for his home, gasoline for his car, food for his table, and furnishings for his house, he wonders what happened to his environment. If the narrow view of the environment is stressed, as is often the case, a bigger environmental problem may result from overkill. By improving one environment, it is possible to make the general environment worse.

To illustrate how the use of the term has progressed, consider the use of the phrase "the environment in the city" in the years 1940, 1965, and 1975. To the query, "Tell me about the environment in New York City," a person in 1940 might have described the parks, theaters, schools, busy streets, honking of taxi cabs, and morning mist. By 1965, the answer would more likely have stressed the air and water, not the schools and shops, and would have mentioned how dirty each was. By 1975, the term would have broadened again to include man-made works, but only in an undesirable form, such as noise, congestion, unsightly structures, and, in some answers, fast food chains.

If this example is correct, it tells us two other things about the use of the term environment. Once it had a broad meaning, including both natural and man-made influences, but later narrowed to include only natural phenomena. Now it has broadened again somewhat to include some of the man-made aspects. Then too, the stress has shifted from desirable to undesirable effects. The terms "environmental pollution" and "polluted environment" are almost like the old Southern expression "damn Yankee." Many children of the sunny South grew up thinking this was one word. Some in the modern generation may think there is only one kind of environment and only one kind of pollution.

### The Meaning of Pollution

The concept of pollution is even more difficult to handle. Perhaps it seems perfectly clear what the term means, so one does not need to define it. In their book *Pollution, Prices, and Public Policy* (1975), Kneese and Schultze must think so, because they do not pause to define it but proceed to describe it. It may be somewhat like the definition of obscenity to Mr. Justice Stewart, who noted that while he could not define obscenity, he knew it when he saw it (Jacobellis, 1964: 197).

One possible definition of pollution may stress, as a polluting activity, any change from the natural form of an asset. We now know that when we put certain amounts of mercury into a stream, microbial action produces a mercury compound that makes the water unfit for raising fish for human consumption. That action pollutes the stream. We have long considered sewage in water, smoke in the air, and garbage on land as changes in the natural state of the environment that are classified as pollution. But when we put chlorine into water, take silt out of it, remove moisture from the air on a hot muggy day, drain a sea or swamp for farmland, or cut down trees to prepare a site for a home, a hospital, or a brewery, we change the natural state of water, air, or land; but not all the changes appear to be polluting activities.

These examples should highlight the point that changes may be good or bad. In fact, it is a central point of this book that changes are almost always good *and* bad, and that the term "pollution" refers only to the bad portion of the action. But this is only part—often only a small part—of the total picture; and to concentrate on eliminating pollution may result in eliminating more good than bad. The ancient axiom that one man's meat is another man's poison is an example of the difficulty in defining pollution; note the diverse reactions to rock music.

Besides the idea of keeping nature "natural" to avoid pollution, there is a general concept that the environment should be "clean." But this too is fraught with problems. Perhaps the classic example is that water clean enough to drink is not clean enough to put in steam boilers. Though tap water may not rust or corrode our insides, it will destroy boilers. Similarly, water clean enough for the manufacture of steel may not be clean enough for the manufacture of textiles. The presence of manganese has no effect on some of the processes in making steel, but may be a serious problem for rinsing textiles. If we were to "clean up" the ocean, the water still would not be fit to drink; if it were, it would not be a fit habitation for the fish who now live in it. Mankind's salvation would mean fishkind's extinction.

The problem of defining pollution may be like that of defining beauty. People say that beauty lies in the eye of the beholder. We suggest that pollution be used to refer to the condition existing when a person finds that the usefulness of some asset to him has declined. This opens up at least two dimensions that are not normally embraced by the common concept of pollution. If some people want to keep a given wilderness in a state of pristine purity, and are successful in so doing, other people who want to camp there find that they cannot. The usefulness of the site to the campers has declined; it is less useful for their preferred activity. Although the site may be natural, undefiled, clean, and pure, if it is unusable, it is in effect polluted. Alternatively, if campers come to enjoy the wilderness, no matter how clean their campsites, wilderness lovers would call them polluters. The usual ideas associated with pollution, such as defilement, uncleanliness, dirt, and desecration, all imply reduced utility. Our proposed definition points directly to that result, but it is not limited by those conditions. Instead, we shall show that cleanliness may be an all-too-common example of pollution.

The second feature of the definition is that it does not name man as the perpetrator. Nature can be the culprit. This is true in two respects. First, nature pollutes in ways we do not consider pollution. Night comes and we cannot see; hurricanes arrive and sweep away much that is in their path; it is unpleasant to go out in the cold; dead leaves clutter the lawns; and there are destructive floods and tidal waves. We should not ignore the occurrence of volcanic eruptions, earthquakes, locusts and plagues over the centuries. The smog over Donora, Pennsylvania, in 1948 killed seventeen people (Townsend 1950: 184); and the heat wave of July 1966 caused an estimated 6,700 deaths in states east of the Rockies (*New York Times,* August 11, 1971).

Furthermore, nature pollutes in ways we do consider pollution, and far outdoes mankind in important respects. Pecora (1970) estimates that the eruptions of three volcanoes—Krakatoa in 1883, Mount Katmai in Alaska in 1912, and Hekla in Iceland in 1947—put more particulates and combined gases into the air than all of man's activities since he has been roaming upright (p. 22). McKetta (1974) points out that while we are spending billions of dollars to reduce emissions of carbon monoxide from automobiles, ninety-three percent of the carbon monoxide output comes from natural plant activity. There are sometimes good reasons to explain why actions by man receive the most attention, but there is also good reason to place those actions in proper perspective; in most cases, the human activities pale by comparison.

The terms "environment" and "pollution" cause problems. By thinking of environment as simply air, water, sight, and sound, we fail to consider food, clothing, and shelter. Improving one environment may damage another. Smokeless air that is also heatless may warm our hearts but not our hearths or bodies. Reluctantly, we shall adopt in our discussion the popular meaning of the word "environment" and use the term to refer to those natural surroundings just listed. But in our analysis, we shall remind the reader constantly of the broader considerations. Our use of the term "pollution" will also be the popular one, even though this use tends to mislead. We shall alert the reader as we misuse the term, and we shall try to broaden its meaning as we progress.

## Man in the Scheme of Things

The discussion so far has been concerned with maximizing man's welfare from his environment, defined both narrowly and broadly. Some people argue, however, that such a goal is an example of man's egotism shining through. Do not fish have rights? Is it desirable for humanity to satisfy its whims at the expense of the beasts of the field and the birds of the air? Or, as Christopher Stone has so well put it, "Should Trees Have Standing?" (to sue). If only man matters, may he not decide that animals should be killed, birds displaced, fish caught, and trees cut, so that he can be happy? All the while, the animals, birds, fish, and trees find their suffering assigned a value of zero in The Great Scheme of Things.

There are at least three ways of dealing with the problem. First, Judeo-Christian theology holds that after God created the world and everything in it, He proceeded to give man dominion over all creatures therein. This dominion evidently translated into food for man: "And to every beast of the earth, and to every bird of the air, and to everything that creeps on the earth, everything that has the breath of life, I have given every green plant for food" (*Genesis* 1:30). "Every moving thing that lives shall be food for you; and as I gave you the green plants, I give you everything" (*Genesis* 9:3). Other instructions given to man were to till the soil and to build homes and fires, all of which indicate that the welfare of man was the paramount consideration in running the world.

A second approach to the problem is to promote the wellbeing of the animals and plants that comprise the environment. This is easy, so long as there is no problem of scarcity of resources. One may maintain a wildlife sanctuary in the Amazon jungle with little or no difficulty if no one prefers an alternative use. The problems arise when it becomes necessary to choose between eagles and lambs; crocodiles and people; and the continued health and welfare of whooping cranes, alligators, rattlesnakes, mosquitoes, houseflies, and smallpox germs, and man, who is also one of nature's creatures. Do we truly want man to sit on the sidelines, talk to the animals, and await the outcome?

This leads to a third solution: let man decide. Indeed, there is no other reasonable alternative. The question of whether or not man will make the decisions about the rights of flora and fauna is analogous to another divertive and divisive question. In resolving disputes, shall we recognize human rights or property rights? The phrase shows the power of the pen, if not of logic. However, regardless of how it is phrased, the question basically concerns whether one party or another shall have the right to use a particular piece of property the way he wishes. When a landlord evicts a tenant for nonpayment of rent, the question is not whether the house shall prevail over the human but whether it will be the landlord or the tenant who shall have the right to use the house as he wishes. The appropriate question is whether we shall recognize the rights claimed by one human being over the rights claimed by another.

So, with our feathered, furred, finned, and flowered friends, any decision about their extinction, survival, or increase will be made by men. In consequence, the question will not pit man against plants and animals, but man against man. Some people may wish to see redwood trees cut to provide housing, furniture, or decorative furnishings, while others may wish to see trees standing to provide shade, picturesque campsites, or a home for nature's creatures. In either case, however, the conflict is between people, and people will resolve the question.

It should be apparent that questions involving the environment also involve choice. For example, should rivers be used to provide a swimming hole or sewage disposal? The popular solution is that we should have the best environment possible and that pollution should be avoided; the river should be used for swimming. An equally supportable position is that we should have the best environment possible and that pollution should be avoided; the river should be used for sewage disposal. Now, the problem becomes one of scarcity—the single stream and the dual demands. Somehow, we must communicate scarcity.

**Notes**

1. Mills and Peterson point out that when the Council on Environmental Quality was formed in 1970, none of the three members appointed was an economist. Among members of its staff, economists have always been a tiny minority. "An economist is struck by the paucity of economic input at CEQ" (1975: 259).

2. From Oxford English Dictionary, Oxford University Press.

3. By permission. From Webster's New International Dictionary, Second Edition, © 1959 and Webster's Third New International Dictionary © 1976 by G. & C. Merriam Co., Publishers of the Merriam-Webster Dictionaries.

# 2 Communicating Scarcity

## Why Communicate?

"In an age of resource scarcities and physical constraints, we are going to have to be a lot more choosey than we have been in the past" (Train 1975: 1). Are we eavesdropping on Adam and Eve during a posteviction conversation? No. These timeless words were spoken in 1975 by Russell Train, Administrator of the Environmental Protection Agency. He has recognized scarcity and is making an effort to communicate the problem.

The need to spread the word about scarcity is obvious.[1] At stake is survival for many—in the long run, for all. But there is more. At what level of happiness shall mankind survive? It is not a new problem, by any means. But words like the quotation above cause one to think that scarcity has just been recognized. Train is not alone. Scholars in many fields are busy at work, trying to cope with the "age of scarcity." What a turnaround! In just 1958, John Kenneth Galbraith added a new catchword to the vocabulary when he described *The Affluent Society*. The tables have turned.

Some economists have become captivated by the hue and cry. This seems almost paradoxical, for as Heinz Kohler has said, economics is the science of scarcity. Few would disagree. For the science to exist there has to be a problem of scarce resources, and the science has been alive and well for at least two centuries. Scarcity of resources has been apparent and the number of scarce resources has been growing for several millennia. Even so, a recent conference had as its title "The Economics of Scarcity."[2]

What explains this sudden focus on scarcity, the need to communicate a newfound reality? Has mankind really been unconstrained in its consumption of goods until now? Were not prices, costs, and budget constraints indicators of scarcity? Have we failed to communicate scarcity?

Perhaps the method of communication has something to do with the problem. When people are admonished about their so-called waste of resources, is that enough to cause them to change their ways? Will a president's urging that people turn down their thermostats truly result in less consumption of energy? Will consumers put bricks in the tanks of their toilets to conserve scarce water, or voluntarily reduce their driving speeds, eliminate vacations, call off the trip to grandmother's house, and not take an ambulance to the hospital in an effort to conserve petroleum? Will swimmers and industrialists reduce their use of rivers to conserve nature's splendor?

To many people, these are bad messages, messages that fail to jibe with their perceived opportunities as indicated by the prices they pay. To others, rising prices are worse news. Is there some way to reconcile these differences?

Prices, wages, and costs are effective messages: after all, they cannot be ignored. If you don't pay, you don't enjoy. It's that simple. But accepting the market as a messenger of scarcity is *not* simple. Many people would postpone the harsh realities of life to go on consuming, somehow not caring about, or at least oblivious to, future implications. For example, the political majority in the United States has supported controls on the price of natural gas. As a result, scarcity has been converted into a shortage. Thus we have snatched defeat from the jaws of victory. The quantity supplied is less than the quantity demanded at the controlled price. But beyond the shortage is a deeper problem. Those who get the artificially cheap product will keep their homes warmer, wear out their furnaces sooner, use decorative gaslights, build second homes, move to colder climates, and use a host of other scarce resources more quickly than they would otherwise. Converting an unpleasantly high price to an unrealistic low price sounds good at first, but what about later?

As illustrated with the word "pollution," communication depends on the perception of the receiver. Given the same message, there can be as many reactions as there are listeners. Because the message of scarcity can be carried in the form of prices through the market vehicle, we must examine reactions to the market. They may provide insight into the problem. When the price of a product rises, the change may generate at least two reactions. Some people accept the market signal as information. The news may be good or bad, but the market is the messenger and is not confused with the message. On the other hand, some buyers hear about high prices and blame the market. Certain sellers feel the same way about low prices. The messenger is confused with the message. Thus, some hope to eliminate scarcity, at least for themselves, by eliminating the market.

**Two Views of Communication**

The two reactions to the market suggest a classification system described by Robert Pirsig. In his writing (*Zen and the Art of Motorcycle Maintenance,* 1974), he chose to divide human understanding into two categories—classical and romantic. Of course, it would be a gross oversimplification of human nature to suggest that particular people are always in one camp or the other. In most cases, individuals are mixtures of the two, depending upon the problem at hand. However, in our discussion, we describe as classical those who accept the market as a messenger of scarcity. Those who view the market as scarcity are termed romantic. The classical economists searched for ways to describe efficiency in the allocation of resources. The goal was human happiness. The market was described as a mechanism that could deliver to mankind higher levels of living.

Pirsig's words describe two views of reality (p. 6).

A classical understanding sees the world primarily as underlying form itself. A romantic understanding sees it primarily in terms of immediate appearance.

He goes on to describe conflict between the two views:

This is the source of the trouble. Persons tend to think and feel exclusively in one mode or the other and in doing so tend to misunderstand and underestimate what the other mode is all about.... And so in recent times we have seen a huge split develop between a classic culture and a romantic counterculture—two worlds growingly alienated and hateful toward each other with everyone wondering if it will always be this way, a house divided against itself.

These two viewpoints become clearer as scarcity is brought into focus. We want to maximize the total environment of man, and in the endeavor, every scarce resource is important. Recognizing the real dimensions of the scarcity problem and communicating these values with accuracy is a role for the market; that is its function. Thus, we shall describe how a smoothly functioning market has dealt with old scarcities. Then we discuss new scarcities. But the fundamental problem of scarcity remains alive and well and living everywhere on Earth. There are conditions that man must endure. How we manage and communicate scarcity is and always has been *the* social problem.

**The Old Problems**

Any resource that has a price is scarce. What could be more obvious? We shall refer to these things as "old scarcities." Almost every source of applied energy is priced. Petroleum has a much talked about price. But that is not new. It was priced when it was first pumped from the ground in Titusville, Pennsylvania, in 1859. And even with all the public concern today about energy, the price of gasoline has risen just now to what it was in 1950, in real terms. Its relative scarcity has changed many times, but with each change society has been informed.

In general, we can say that the categories of goods called food, clothing, shelter, and transportation are old scarcities. The things that once were called "necessities of life" are priced. They, too, are old scarcities. There are millions of goods that have had prices for centuries. Their scarcity is obvious.

Although the idea of price and old scarcity sounds simple, it is not. The development of markets with complex systems of property rights, contracts, money, and information flows took centuries. In the simplest form possible, efficient markets require five active participants. As John R. Commons (1957) pointed out, first there are the two parties who wish to trade with one another. In order for them to gauge the value of their goods, there must be an active buyer and seller of some substitute item. Finally, there must be a referee in the transactions, someone to enforce contracts and protect property rights. But once the bidding starts, the market communicates hopes, dreams, and scarcity. The total environment of the active participants becomes captured in the market process. Price is the signal.

If communicating scarcity is a problem, it has an obvious solution. Use the market. It is not a new institution. Markets are very much alive. Prices are everywhere. So what is the problem? The problem seems to relate to the classical and romantic views of things. There are people who question the parity of the market place. Some people are concerned about equality on the one hand, efficiency on the other. Sometimes the former gets in the way of the latter. Or vice versa.

**The Tradeoffs: Equality Versus Efficiency**

As we have said, if communication of scarcity is important, a smoothly functioning market will spread the word. Prices and costs provide unambiguous incentives for placing resources into their most productive uses. Following their own self-interests, individuals will make exchanges that result in improving social welfare. That is the market game. Happiness is the result. And the ability to be efficient affects each person's outcome. The outcomes will not likely be equal. However, the openness of the market will give equal access to everyone.

The parity of the market is such that each person is measured by his ability to please others—his productivity. The extent to which each individual participates will be limited by his endowment of goods, his wealth, human and otherwise. Even so, there are costs that weigh in the balance, costs that fall on the rich and the poor. A person with wealth has more to conserve, more to lose. Thus, if some people begin with goods, they may end in bankruptcy. Perhaps they are inefficient players.

For other people, the reverse holds true: they begin in relative poverty yet end with wealth. By keeping the market process open, all individuals have the freedom to succeed or to fail. The process generates income; it also distributes wealth. There are no guarantees. When the market signals of scarcity are observed, response to those indicators of social need and want can lead to success. Failure to heed society's beck and call leads eventually to failure in the market. The market communicates and excommunicates. It is not perfect. However, what is?

The operation of the market can be thought of as very impersonal, but it has the deepest of social implications. At its base are the preferences, the demands, the very souls of people—people who show their willingness to trade off one scarce resource for another. Its implications are real, not imaginary.

With this brief introduction, consider Pirsig's classification system. How might a romantic view this process? Prices rise and fall; individuals move from one economic role to another; wealth moves from the unproductive to the productive. Without seeing the market as an information system, an abstraction of real conditions, the romantic might suggest that the whole process be stopped. Or if it is only a game, change the rules, just slightly. Make sure that no one gets hurt in the process of producing happiness. So, the romantic may call for a limit to the domain of inequality in the market by recommending that certain activities be controlled by external, nonmarket forces.[3] That is, a forced redistribution of income can be imposed on the market. Or attempts to achieve specific levels of consumption for all parties can be imposed. The romantic will call for laws and regulations, as he confuses the market with scarcity.

This is more than just a possibility. It happens with unending frequency. Even so, agreement on those activities that should be paid for by other people is difficult to obtain. Pondering the possibilities raises almost impossible questions. For example, if a minimum level of living is to be assured, just what should that minimum be? Will it be the same for each family, each individual, without regard to preferences or even to location? What is a minimum level of housing in the humid southeastern United States? Is it the same as in Alaska? What about medical care, education, nutrition, and clothing? What about energy? Where does the list of necessities, minimum levels of living, end?

A modern Solomon could not unravel the puzzle. Yet answers are given. For every intervention in the market, there will be a change in the message system. In effect, minimum levels of living assured by social action sometimes introduce zero prices on certain items. The communication link between scarcity and human action is cut. Bad news changes to the best news. An item is free, and according to the law of demand, large quantities are demanded. As a consequence, the prices of other goods change so that more of the socially ordained goods can be produced. So, other prices tend to rise. The messages from the market become filled with static. Scarcity becomes disguised.

Those classical persons whom we mentioned earlier tend to counter the process of artificial price reduction by the romantic. If one views the market as an information system, a means of communicating scarcity, one seeks to have the information as accurate as possible, not as pleasant as possible. To the classical individual, accuracy is good. Too much is at stake to alter the messages. A misunderstanding can lead to irretrievable losses.

Looking at the elements of the market, the classic may observe the failure of a firm. This becomes a clue of inefficiency. Managers are supposed to learn what consumers will buy and to produce those goods as cheaply as possible. The market forces this behavior. By competing for the use of scarce resources as indicated by price, by striving to make profits, managers learn. The word from society comes through in a very direct and forceful way. In a similar way, owners of resources are expected to do the best they can in the market. The price signal is again the message. By accepting the highest price possible for labor, land, or machinery, the owner moves the resource into its most productive role.

To disguise price in all this would subvert the entire process. Classics consider such action similar to recalibrating thermometers to ward off winter. To set price at zero is to break the thermometer: the message is never bad, but it is far from accurate.

Such diverse views of scarcity can be viewed in another way.[4] Perhaps classics see price as a variable, a signal reflecting relative scarcity. The classic is interested in the real resources as they are allocated by price. Scarce property rights are his concern. The romantic may see price itself as real. Property rights to price (that is, control over the market mechanism) is seen as a way to gain control over resources. Of course, if someone can control the market, the individual in command can pick and choose the characteristics that are most beautiful to him.

The presence of these two views leads to conflicts. It is clear that a stage has been set for the disputes that rage. It is the political arena, for the political mechanism sets the rules of property, the rules of the market. Furthermore, government has the exclusive power to tax, to regulate, and to redistribute income on a collective basis. Here the conflicts come into sharp focus.

### Newly-Recognized Scarcities

Though classics and romantics will continue to struggle over old scarcities (that is, those resources that already have prices), there is another problem—goods that have not been priced. What about the new goods? What are they? The term "environmental quality" says a lot in answer to the questions. Air and water quality, scenic views, sounds, smell, justice, safety, all of these and more are now recognized as scarce. The tons of paper, hours of television and radio time, months of legislative hearings, and years of academic discussion bear testimony to efforts to communicate about new scarcities. As yet, however, that most sophisticated mechanism for communicating the problem—the market—has been used sparsely.

Perhaps reluctance to use the market reflects a victory of the romantic over the classic. If people view the market as the creator of scarcity, it follows that they will develop some other mechanism to deal with these problems. But, there is more.

**Property Rights**

For a market to work with full efficiency, it is necessary that property rights be established to the new scarcities. This is not a once-and-for-all process. The new scarcities are, to a great extent, characteristics of old goods. Old scarcities such as water and land have always had certain qualities associated with them. For example, are the rights to the sun's rays embodied in the land that receives the solar energy? Or is it possible for an adjacent land owner to collect that energy or by some process to deflect the rays away from the property in question? Are the rights to neutral odors associated with the lease of a particular apartment?

The questions posed relate to land, one of the more visible scarce resources. However, the characteristics of space are almost without limit. In "Some Basics in Land-Use Policy," (p. 7), Gene Wunderlich gets to the heart of this problem when he poses the question, "What is land?" His answer makes the following point:

*Land is space.* It is two, three, or four dimensional space. It is the boundaries within which all else is stated. It is more than measured; it *is* the measure. Land is front-back, left-right; it is up-down with minerals, sunshine, energy, warmth, sky, trees, slopes, soil, and moisture. And land is change, seasons, growth, and evolution.

Several qualitative aspects of land are suggested. Each one could be plentiful or scarce, relative to the desire for it.

We could go on with a discussion of silence, and other qualitative aspects of life. But the point is this: these serve as examples of the new scarcities that have not been made explicit parts of the property rights system. They are not marketed directly. There will be problems in dealing with these new scarcities unless the rights to them are clearly defined, made transferable, and give to the owner the full use of benefits from the new asset.

The process by which full property rights develop and become allocated by markets is a lengthy and continuing one.[5] As pointed out here, certain characteristics, once plentiful enough to be free, sometimes become scarce. Controversy results over their use.[6] Conversely, some characteristics once scarce, become plentiful. As price falls, enforcement of those rights may be reduced. The controversy over use disappears.

## A Hierarchy of Rights

After we recognize that property rights to certain characteristics give value to real resources, we must identify the process by which these rights are defined. This is discussed later in some detail. Simply classified, there are three levels of property rights to scarce resources: common, public, and private. Within these, there are various schemes that can alter social behavior.[7] In other words, there is much more to a property rights system than the mere selection of one of these levels from the hierarchy of rights.

When people confront a new scarcity for the first time, it must be in a setting of common property. Everyone has had access to the resource. Each individual decides whether or not to participate in the use or consumption of the resource. No one can exclude another person from access. Fishing and hunting are common examples of this case, but there are more—many more. Water quality, not quantity; air quality, noise, odor, solar energy, weather, and certain land uses are also examples. However, along the way something has happened on the "commons": the common resource becomes crowded. Individuals begin to sense the new scarcity and to realize that although they are not able to exclude others, they are affected by the entry of others. Predictably, this recognition brings action; a transition occurs in the property hierarchy. An example may shed some light on the process of transition.

In the early years of radio broadcast, it became evident that individual broadcasters could no longer enjoy unhampered use of the airwaves.[8] People voiced an appeal for help. Action came from government; the Federal Radio Commission was formed. Thus, *common property* became *public property*. Several people decided who would and would not use the scarce resource. Exclusion became a public power. The same transition has occurred with other resources. Zoning ordinances, effluent discharge permits, air quality standards, safety standards, offshore drilling rights, and controls for aircraft landings—each involves a decision made by public bodies. These rights to use are public property.

Under a system of public property, a small group of people decides what will be done with property. The resource in question is withheld from some of the public, and given to others. Some political processes give the public access to the decision making. Zoning commissions, environmental control agencies, railroad commissions, and public service commissions are examples. Public hearings are an attempt to read the wishes of the common owners. Conflict, sometimes bribery, crops up. The political process itself now has value. In a sense, it takes on a price.

But the price signals are not clear. The rights to use public property have an apparent price of zero. One cannot buy the right to broadcast; yet, the right has great value. So, it pays an interested party to act strategically in an effort to gain control of some part of the resource. Scarcity is now obvious, but there is a shortage of information about the *extent* of scarcity. Not all interested parties are able to communicate effectively. Only people highly experienced in communicating with commissions and boards will be effective in obtaining use of the scarce resource. They themselves will be scarce and command a high price.

The decision to communicate scarcity by moving to a system of public ownership is a choice. There is an alternative—*private property*. However, this is no simple move; it requires caution.

For example, when people realized that air waves were scarce, the government, after appropriating the rights, could have refereed the process on a once-and-for-all basis. Several approaches were available. The rights could have been allocated initially as they were; that is, they could have been given to the most "desirable" parties. The gift could have been defined and given in fee simple. The new owners would have been able to sell or to use the rights. Restrictions could have been a part of defining the asset, but a market could have developed. Price would then communicate scarcity. The social value of the new resource would have tended toward its maximum. Owners would have taken care in using the valuable resource. The rights, not good will, would have been an asset.

Another approach was possible. By giving the new asset to the "most desirable" applicant, a deliberate redistribution of wealth takes place. This initial gift could have been avoided. Even those first applicants could have submitted bids with their applications. In either case, however, a market would have developed.

Private-property rights are power in the hands of the owners. Thus, the process through which they arise has social effects. These must be considered. Warren Samuels (1972: 135) has said, "Property rights do count: they enable their owners to visit externalities upon others to begin with." That is, rights enable the holder to affect other people by withholding his property. That is what property is all about. How one obtains the right, the way in which this power is created and dispersed, is an important question.

To avoid arbitrary distributional effects, a neutral third party must appropriate newly recognized scarce resources, define property rights to the resource, and auction the rights. Since it is asserted here that such resources were common property first, it may make sense to capture the initial value of the new private property for the commonwealth. Government is the logical neutral third party or referee. Once the initial rights are defined and sold, the revenue can go into the common purse; taxes can be reduced.[9] After the sale, the government becomes simply an enforcer of contracts, offering protection for private property.[10]

The inclination to bid for scarce rights depends upon anticipated income from the new asset. The possibility of bidding introduces a cost, a cost for holding and withholding property. Because every person has access to the bidding opportunity, the resulting ownership pattern will reflect social choice. Power will disperse from a centralized body politic and diffuse into the market. Though that feature may hold interest for both political and economic theorists, there is another interesting aspect. The cost of holding property *communicates scarcity*. Power has been established, but at a price. Each owner must pay to withhold a resource from others. Access to the market insures that those who value property the most will obtain the right to use it. This presents an interesting paradox of sorts.[11] For markets to function properly in allocating private property, the markets must be common property. If markets become private property, affecting exclusion, then the social function of price is distorted. Markets are a part of the common wealth. Property rights are private wealth.

### Rights to Communicate

By defining private rights to scarce assets, individuals in society obtain the right to communicate scarcity.[12] The common ownership of markets guarantees this. Property is a social process, a social invention allowed, indeed encouraged, by society in order to communicate and to maintain life.

The economic right to communicate introduces powerful psychological forces. Prices and costs require that the preferences of others be recognized; this recognition places a limit on the behavior of each individual. It has been called "mutual coercion."[13] Without price, any "owner" would carry the use of a resource to the limit of his will alone. Forbearance would be the only limit to the use of newly recognized scarce resources (Commons 1957: 77).

Recognition of individual willpower causes some to plead for a new religion or ethic to deal with newly recognized scarcities. Scarcity becomes a moral issue, an ethical question, calling on the fragile forbearance of individuals to limit the use of society's scarce and valuable resources. Questioning the proper role of markets, Arthur Okun (1975: 119) discusses the advantage of the market on this point:

Most important, the prizes in the marketplace provide the incentive for work effort and productive contribution. In their absence, society would thrash about for alternative incentives—some unreliable, like altruism; some perilous, like collective loyalty; some intolerable, like coercion or oppression. Conceivably, the nation might instead stop caring about achievement itself and hence about incentives for effort; in that event, the living standards of the lowly would fall along with those of the mighty.[14]

The effects of pleas for new values are not to be discounted, however. People do reflect their values in all economic activity—priced and otherwise. The market process reinforces any tendency which people have to forbear. Of course, there will be some true believers who feel that total abstinence is the only acceptable use of certain resources. Some romantics may advocate zero waste discharged in streams, zero oil purchased from Arab sources, and perfect purity of air, sight, sound, and safety— even no billboards. To them, the only acceptable price is one that limits to zero some particular consumption or use of certain unique assets. These advocates of purity would force their will on others, acting in effect as if they had the property right and hence the power to withhold resources from others. They would monopolize the rights to communicate scarcity, and markets would be transformed from common property to private property.

**In the End, Scarcity**

Whether it be new or old scarcity, we are left with it. Failure to recognize this reality is utopian. From this standpoint, the prospect of eradicating scarcity is not very bright. But there are options as to how we live with scarcity. We can communicate it effectively and, by doing that, make the best of a bad situation.

The crisis mentality surrounding the "new" age of scarcity works against our dealing with the problem. There is constant confrontation with the totality of each problem. Instead of focusing on necessary adjustments, people raise questions about the *total* world reserve of petroleum, the *total* stock of natural gas, and the *complete* dimension of the ozone shield. Such questions force one to feel desperate, to think that the total problem must be solved immediately. In addition, they lead to erroneous economic questions. What is the value of the environment, all of it? Or of the ozone layer? Or of the ability of ten states to have energy? The dimensions of the questions are too massive to be comprehended.[15] One is inclined to answer these questions with equally massive numbers.

There are arguments, even laws, that would devote a nation's annual economic product toward the protection of just one part of the environment, such as water. That is not the trade-off to consider.

Throughout history, only the edge of scarcity has been communicated. In a similar way, only the edge of human effort has been devoted to solving problems. That is enough. What is the value of one more generating plant? One more waste treatment plant? By thinking in terms of marginal analysis, people can make correct decisions. Individuals can make judgments that alter activities by degree rather than in totality. The trade-offs are at the margin.

**The Growing Edge**

The growing edge of scarcity confronts the growing edge of society. Communication takes place between the two. The more information channels there are within society, the greater the message flow. Markets, prices, and costs link every person in a market economy. Thus, the growing edge of a market society is very sensitive to change. *The market is a medium of change.*[16]

Of course, not all people stand at the outer edge of the system. There are many edges, many levels of living, many aspirations. Some people live the way their grandparents did. Some maintain only a minimum level of subsistence, barely hanging on to fragile threads of survival. Some do not survive, while others move quickly through various levels of affluence. In a market society, there is mobility upward and downward.

The growing edge of the social system is characterized by a flood of information. Those at the most highly developed edge see, hear, read, and experience much of the world's events. The computer-bred information revolution guarantees that by pressing a myriad of local, regional, national, and world problems upon their consciousness.

As a result, many of the informed individuals have the best of everything—the most affluent living standards, the best education, and the best information. Yet some on the edge claim there is no hope for man. Why? They constantly look back, moralizing about the ethics of success and failure, raising questions of equity, asking, "Is it right that we at the edge should have so much, know so much, and the others...?" Sometimes they conclude, "We should stop producing, using up the world's resources, eating so much; our success has capsized the world." Thus, some classics become romantics and pull away from the growing edge of society, the very edge that confronts the most stubborn scarcities.[17]

There is some irony here. The reflections about the misfortunes of man outlined above usually are propounded in very comfortable quarters by individuals who expect to live out their lives in relative abundance

(Jukes, 1973: 11–12). Yet the dilemma they force on themselves tends to retard society's growth. Most people who support no growth have already secured a level of comfort sufficient to pass some on to their next generation. Perhaps they have a feeling of glee for a declining G.N.P.

Farther away from the growing edge of society is the multitude slowly moving through early stages of economic development. There, the first and most severe battle with scarcity is raging. The multitude understands absolute scarcity. It is real.

Whether at a subsistence level or at the very limit of abundance, there is information to be communicated. Decisions must be made. Plans must be made. Eating less, working less, producing less will lead to less for everyone. That is no way to communicate the scarcity of resources. Commands, regulations, ethical promptings may illuminate the problem but, as a concomitant, will limit choice. An efficient market can communicate. It can allocate and distribute. It can accommodate growth and change. The use of the market for communicating scarcities—old and new—is an option that deserves careful consideration by romantics and classics. By strengthening the market mechanism, it may be possible to achieve higher levels of happiness for mankind. And that is the name of the game.

We turn now to apply these principles to the natural environment.


**Notes**

1. John Z. Young (pp. 157–162) stresses this point. He suggests that life scientists will eventually join forces with their colleagues in the social sciences to address ultimate social questions. He goes on, "The aim of the new unified science might be said to be to define those relationships between populations of people that enable them to communicate information and so to maintain life." He warns, "To communicate is not our whole nature. It is our means of getting a living as a social animal, but it is only a means, not living itself."

2. This particular conference for economists was held at City College of New York in May, 1975. Another example of the new focus is seen in the theme of the 1975 meeting of the American Academy of Political and Social Science, "Adjusting to Scarcity."

3. For example, see the writings of Tobin (1970), Heilbroner (1974), and Okun (1975). Senator Hubert Humphrey calls for a more strenuous approach, suggesting that an economics of abundance can replace the prevailing economics of scarcity.

4. The conflicts are presented in another way by Bruce Yandle in "Property in Price" (1975: 501–514). Also see McKean (1972).

5. Gene Wunderlich discusses the process with reference to land.

But an exhaustive discussion of law and markets is found in the classic work by Commons (1957). Two recent collections of readings, one by Furubotn and Pejovich (1974) and the other by Manne (1975), deserve attention.

6. This controversy is discussed in an interesting way by the United States Council on Environmental Quality. Under the subtitle "The Basic Values to be Balanced—Private Property and Public Environmental Concerns," which itself is telling, follows this statement of the problem: "It has never been the law, of course, that title to land confers the right to use the land however one pleases. . . . Property rights, in short, do not exist independently of the protections and responsibilities linked with such rights by the law. *As those legal protections and responsibilities change to reflect new perceptions of society's well-being, so also does the concept of private property*" (pp. 125–126). Italics added.

7. Tideman (1972) sets forth these three systems in a brief but clear fashion. An analysis by Pejovich (1974) gives details of various schemes that can be introduced in one system. To a certain extent, there are rights within a broader system. For example, certain East European economies hold capital to be public property. However, there are a variety of private rights which can flow from public capital. These may be attenuated by the state (Pejovich 1974).

8. An interesting discussion of this is found in Wilcox (1971: 536–538).

9. A more complete discussion of this is given in Yandle and Barnett (1974), and Barnett and Yandle (1973).

10. Schmid (1967) touches on the government's role as an appropriator of newly recognized scarce resources and its later function.

11. See Dales (1972: 154).

12. It should be recognized that markets do not obtain without costs. Thus, the right to communicate through price does not come free. Harold Demsetz gives an interesting discussion of this cost and points out that when goods are scarce, "we need information which is obtained by excluding nonpurchasers, *provided that the additional information is worth more than the exchange and police costs necessitated.*" See Demstez (1964: 20).

13. Samuels (1972) defines the term "mutual coercion" and enlarges on it. He draws upon the earlier work by Commons (1957).

14. Heilbroner (1974: chapter 5) also reviews the prospect of ethical change as a means of meeting scarcity's challenge. He sees little hope for success on this score.

15. The old saw, "Economists know the price of everything, the value of nothing" is mentioned by Schmid (1967). His point is that mentioned above: no one can possibly know or estimate the value of total resources.

16. This idea was suggested to us by Dr. Russell Shannon. It deserves further development.

17. John Maddox (1972) addresses this point: "Still more serious is the danger that the extreme wing of the environmental movement may inhibit communities of all kinds from making the fullest use of the technical means which exist for improvement of the human condition" (p. 32). Going on, he says, "Just as it gives offense to tell developing nations not to build blast furnaces that pollute the air, so it gives offense to suggest to relatively poor sections of the community that the time has come for the search for material prosperity to cease" (p. 279). Here again is the problem of communicating scarcity. It is not necessary to "tell" nations or communities anything if the price system is developed to its full capacity. Price tells them.

# 3

## The Environment As Scarcity

**Congestion and Scarcity**

Scarcity has characterized man's existence and has constantly command-ed his attention. Finding happiness in a world of scarce resources has been the driving force that pushed for solutions to the problem. Where scarcity existed but appeared to be beyond solution—bringing rain to the deserts, or drying out a monsoon-drenched jungle—man concluded that his time would be better spent elsewhere. Some problems were endured, others were reduced. Relative costs and the happiness pay-off prescribed the course of action. Thus, the market communicated and became the means for dealing with the problem of scarcity.

Industrialization has made the problem more complex and life more tolerable, but the principles of scarcity are the same. When the market solved the modern problem of allocating efficiently a product like steel, another problem remained. Smoke and dust were produced when the steel was produced. Some people got clean steel shelves, but others got dirty linen. And there seemed to be no way for the two groups to communicate.

To the extent that dirty linen represented a cost, something was missing when decisions were made about producing more steel. It was an *externality*, a cost not included in the price of steel. Economists wrestled with the problem, but generally they lost. A. C. Pigou (1920: 193) cast the problem in terms of highway users, pointing out that drivers would crowd onto a highway, causing delays for all drivers in the stream of traffic. Putting another car on the road was analogous to putting more smoke in the air. Additional cars and smoke may bring more cost to others.

Pigou reasoned that the problem from cars arose because of the profit-seeking motives of the owners of vehicles. In pursuit of their own happi-ness, they reduced that of some other highway user. He proposed a charge on all users, once crowding occurred. The helping hand of the toll gate operator would give aid to the drivers. Less congestion and more rapid movement of the paying customers would result. Those who valued the road less than their money would be free to make other, less costly arrangements. Yet, somehow the need for a charge was viewed as a failure of the market.

Frank Knight (1952: 160–168) saw the same problem in a different light. Users would tend to crowd the highway, but the problem of congestion would arise because of a lack, rather than a surfeit, of self-interest. By failing to charge cars and trucks for using the highway, the owner of the highway did not maximize his profit but did help to create the congestion. Knight's solution was not that the government charge to offset the greed of drivers, but that the government charge to evidence its own greed. By seeking to maximize its revenue from tolls, the government, as the owner of the highway, would simultaneously maximize the welfare of society derived from the use of the highway.

A useful rule in dealing with any problem involving the use of a free asset is to couch the same problem in terms of something more familiar, something for which there is charge. Consider land. Instead of many cars seeking to enter a particularly convenient highway, assume there are many farmers seeking to farm a particularly fertile piece of land. Ricardian rent dealt with this problem 150 years ago. The theory held that the owner of the most productive land should charge a rent for its use to prevent overuse and, therefore, inefficiency. At the same time, the owner would maximize his profits. If the owner of the freeway acted like the owner of the land, there would be net gains to both owner and users, although not every user could claim a gain.[1]

For example, assume Jones can go from A to B in one of two ways. He can take the long way around, which takes twenty minutes, or he can take the expressway, which takes fifteen minutes. However, if he takes the expressway, he slows down traffic so that each of the one hundred other cars already using the road takes fifteen seconds longer to make the trip. In sum, Jones saves five minutes by using the expressway, but the other hundred users lose a total of twenty-five minutes, adding that much to their travel time. If time is equally valuable to all highway users, the sum that other users would pay for what they consider improved service—that is, a faster trip without Jones on the road—would be more than Jones would pay to enter. Only if the value of his time saved exceeds that of the others would Jones get to use the road. The same condition applies to every other user of the road.

Knight published his reply to Pigou in 1924, but it was almost forty years before any significant advance developed in the analysis of the problem. Even Knight's example did not clearly involve externalities. Like Pigou, he dealt with many people trying to make the same use of one asset, a highway. The problem was congestion. It still seemed that smoke was a different problem. With congestion, Jones crowded Smith, and vice versa. With smoke, Jones creates and Smith suffocates, but not vice versa.

In 1960, Ronald Coase linked the economic dimensions of congestion and pollution in his seminal article, "The Problem of Social Cost." Coase cut to the bones of the problem and found that supply and demand—the market—explained the situation. Rather than focus on cars, steel, or smoke, Coase chose to emphasize cost. When one person demands more of a scarce asset, this places a cost on another demander. If ownership to the asset is established, transactions between demanders can reduce the conflict. The externality disappears.

Coase's idea that costs and externalities are bilateral was startling. The idea was seized upon, cited by many, but understood by few. The central concept is that when the mill's smoke bothers its neighbor, Mrs. Murphy, as she hangs out her clothes to dry, both parties are the cause of the problem. It was easy to see that when the mill put out its smoke, that hurt Mrs. Murphy. It was less easy to see that when Mrs. Murphy put out her clothes, she hurt the mill. In fact, she did not hurt the mill. There was another condition needed. If the mill got its way and put out the smoke, Mrs. Murphy was hurt. If Mrs. Murphy got her way and put out her clothes in government-ordered clean air, then the mill was hurt by having to put out its fire or put in a precipitator. But many people believed that Mrs. Murphy was not causing a problem: the mill was doing only what it naturally should have been required to do.

The logic of the bilateral argument was faultless, but it had no heart. Few people were willing to argue for dirt, even though its production might also create many goods. And who could argue against clean clothes? or Mrs. Murphy?

**The Modern Theory of External Cost**

In each of the examples considered so far, there has been a similar problem. First, there was a resource: a highway, a clean environment. Then that particular resource, (space, environmental quality) became valuable—more than one person wanted it. Conflict developed over use. All parties were prompted by their self-interests to attempt to use the scarce resource. As a result, they got in each others' way. The more familiar statement of the problem involves pollution.

Let us assume that located near the headwaters of a river stands a mill producing a product, textiles, that gives rise to a waste product, sizing, which is put into the nearby river. Assume further there are a hundred residents who live downstream along the banks of the river and that the sizing in the river bothers or damages them in some way. We may plot the marginal effects on those upstream and downstream when waste is put into the stream as shown in figure 3–1.

The total cost of waste treatment is an exponential function: the more waste that is removed from a given amount of effluent discharged, the

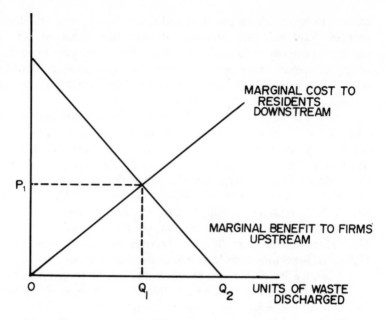

**Figure 3–1.** Marginal Effects of Waste Disposal

more rapidly the cost of waste removal increases. The second derivative is positive. Engineering studies show that it may cost as much, or sometimes more, to remove the final one percent of an undesired waste from a given amount of effluent discharged as it cost originally to remove the first ninety-nine percent.

Though it may exaggerate the relationship in some cases, table 3–1 provided by the Environmental Protection Agency shows how investment costs rise as greater levels of water purity are attained. Note that it costs four times as much to remove the last fifteen percent of waste as it did to remove the first eighty-five percent of waste. It costs fifty times as much per unit of waste removed to go from ninety-nine percent removal to one-hundred percent removal as it does to go from eighty-five percent to ninety-five percent removal. This is no more than a special application of a general rule: the closer one approaches perfection, the harder it is to make any more progress. It helps to remember what we already know. As a consequence, if a firm had been required to remove all of the waste substance from its effluent, at a very great cost, and then were suddenly told it need remove only ninety percent and would be permitted to discharge the remaining ten percent into a nearby waterway, there would be great joy at Mudville Mill that evening.[2] The marginal benefits from discharging the first few units of waste would be high; the benefits from

**Table 3–1**
**Index of Pollution Control Investment Costs Related to Level of Abatement**

| Level of Removal (percent) | Cost of Removal (Indexed for 85% removal = 100) | Increased Cost of Removal | Increased Percent of Removal | Cost per Increased Percent of Removal |
|---|---|---|---|---|
| 100 | 500 | 250 | 1 | 250 |
| 99 | 250 | 50 | 1 | 50 |
| 98 | 200 | 50 | 3 | 17 |
| 95 | 150 | 50 | 10 | 5 |
| 85 | 100 | Indeterminate | Indeterminate | Indeterminate |

Source: Adapted from *The Economics of Clean Water*, 1972, Summary Vol., p. 23.

discharging more units would be less and less. The function is like the marginal benefit to firms upstream shown in figure 3–1.[3]

Similarly, the total damage done by these units of waste put into the stream rises exponentially. The first units discharged may be absorbed or diluted into insignificance; dilution is almost always one solution to pollution.[4] Additional units may make the water murky; more may kill plants; still more may kill fish; and yet more may make the water smell bad all the time. Marginal costs to residents downstream rise as each extra unit of waste discharged adds more discomfort.

However, there is some hope in all this. There is a "best" solution buried within the graph. If the first units discharged save much money but cost little to endure, some use of the stream should be made for that purpose—waste disposal. But if more and more of the waste put into the stream is less and less beneficial, but more and more costly, there comes a time when no more waste should be put into the stream. There is an optimum level of stream use when $Q_1$ units of waste are discharged into the stream. At this point, the marginal benefit from the last unit of waste dischraged is just equal to the marginal cost it creates. Any movement from this point will benefit one party, but at a greater cost to the other.

**Policies for Achieving the Optimum**

Now that we know where to go, the next problem is to devise a policy that will get us there. Three major policies have been suggested: regulations, subsidies, and charges. Each has had its own following. Politicians and the bureaucracy have generally preferred regulations; businessmen have pressed the advantages of subsidies; and economists, in a rare burst of relative unanimity, have praised the use of charges. We shall look at these three policies by assuming there is a question regarding the use of a river. However, the principles involved have far broader application. They apply to air, sound waves, land, and the current questions of abortion, welfare, and freedom of speech, to name only a few.


## Regulations

Regulations are by far the most widely used method for dealing with environmental use. This is true in the United States, in other well-developed countries, and in those that are not so well-developed. The announced advantages are that regulations encourage efficiency by offering a greater degree of certainty, and that they reduce transaction costs. There are other, less publicized reasons that may explain their popularity more accurately. We shall offer an evaluation of each of these arguments.

Under a system of regulations, the government passes laws, issues regulations, or provides rulings that govern the quantities of waste that may be put into rivers and streams. In terms of figure 3–1, the government would allow the upstream firm to discharge $Q_1$ units of waste into the stream. Because it is not popular to state that $Q_1$ units of waste can be discharged, the regulation is more likely to order that $Q_2 - Q_1$ units of waste must be treated and removed by the firm. The public sees positive action taking place; $Q_2 - Q_1$ units of waste are being treated because of government orders. The firm sees a silver lining on this dark cloud because it understands that $Q_1$ units of waste need not be treated and can be put in the stream. The positive way of stating the policy gains support from people who want to enjoy cleanliness in streams.

On a more technical note, there is the announced advantage of certainty. Those who revere regulations observe that if it were not for government rules, firms would not know with assurance how much waste they would have to treat at different times or how much they would have to pay to use the stream for disposing of their wastes. Because of this uncertainty, they would be less able to prepare for the future and to produce efficiently the goods people want. This certainty, however, is more apparent than real. Changes can and do occur. First, Congress may change the laws. It is well known that the Lord giveth and the Lord taketh away, and Congress is not without its own delusions. For example, Congress passed the Federal Water Pollution Control Act in 1948. Eight years passed before it was amended, but since that time, the period of adjustment in water laws has been two to five years. not counting laws dealing with the environment in general. And the adjustments have been significant for business, changing markedly the rules governing the use of water. Laws can render a plant obsolete and useless even more quickly than can technology. Consider plants producing cyclamates, DDT, and chlordane. The *Wall Street Journal* (December 23, 1970: 1) cites a case of an electric generating plant that spent $2 million to meet state air quality emission requirements, then two years later was faced with new regulations that made its new pollution abatement equipment obsolete. There was no way the equipment could be altered to meet the new demands placed on the plant. A new and different system was necessary—at a cost of $4 million. Predicting consumer tastes, technological change, and acts of God is enough to strike fear into the hearts of producers. Trying to predict the acts of Congress certainly belongs on the list when laws and regulations are used to govern business activities. Perhaps it should head the list.

An unusual variation on passing new laws is discovering old laws. This does not appear to be so prevalent, but perhaps the process is still too new. An excellent example concerns the River and Harbor Act of 1899, which included a recodification of certain provisions related to navigation. These provisions become part of the United States Code and were applied with relative consistency until the 1960s. In 1970, these provisions attracted widespread attention; so much so that the Act was resurrected and rechristened The Refuse Act of 1899. This rebirth was more in the Hindu tradition of reincarnation in a totally new body. Whether this later form of.life constituted a movement toward or away from nirvana is a debatable question. The point is that old laws may lie dormant until changed conditions give them a new lease on life, when they become the basis for even greater changes in conditions. For a discussion of this metamorphosis, arguing that it is natural, see Tripp and Hall (1970); for arguments that it is unnatural, see Hite (1972: 85–89).

The third attack on certainty stems from changes in the interpretation of given laws. Like the well-known comment of baseball umpires, "it ain't nothing until I call it," the law is sometimes uncertain until it is interpreted by those given that responsibility: administrators and judges. An unparalleled example of the absence of certainty due to this cause concerns the construction of the Alaska pipeline. While construction firms expected to get underway in early 1970 and shipped equipment and supplies to construction sites at that time, changes in interpretation of the relevant laws by administrators and courts delayed the start of construction until Congress finally took the matter in hand in late 1973 and authorized construction of the pipeline. Just the road-building equipment of five contractors sitting idle during the early part of the delay was said to have cost these firms $100,000 a day (Hoffman 1973: 1610). The State of Alaska estimated its daily loss in royalties and taxes at $750,000 a day in 1971.[5] Few sources of uncertainty can produce a four-year delay in a project and help contribute to an increase in its cost from $1 billion to over $5 billion.

There is one other aspect of uncertainty that deserves mention: how the form of uncertainty affects the efficiency with which it can be overcome. Both prices and laws are forms of uncertainty, but laws leave one less option for the person affected. This has long been recognized in international trade, where the effects of tariffs and quotas are often compared. If a quota is established to forbid the importation of some good like meat, buyers must turn to alternative domestic sources of the good no matter how expensive these sources may be. On the other hand, if instead a tariff is imposed on the importation of the good, buyers may still rely on foreign sources if domestic sources are priced above the foreign price plus the tariff. The same is true for regulations for water quality versus user charges. With laws and regulations governing the use of a stream, the potential user must turn to alternative man-made devices, regardless of cost. Removing the options to pay more and still use the good or service not only limits freedom, it reduces efficiency.

The second main advantage of regulations over alternative forms of control is a predicted reduction in transaction costs. Economists in particular promote this advantage. The classic case deals with residents around an airport who place a value on decreasing the noise from airplanes' taking off and landing. Of course, the airlines may not cherish noise, but they would like to avoid installing noise abatement equipment, as this would raise their costs of operation. Again, figure 3–1 is relevant. The problem arises in contacting each of the affected parties; eliciting from each an honest evaluation of how much he would pay, or charge, for less, or more, noise; collecting from those who would be willing to pay; and paying those who reduce their decibel output. After we get through with all this, we may discover the game was not worth the candle.

If regulations were used instead, the government could estimate or otherwise determine the value that each of the parties places on his use of the environment, compute the optimum solution, and order airlines to reduce their noise accordingly. The cost of all this would indeed be much less than private arrangements. But, there is more to be told.

Transaction costs are often viewed as unfortunate man-made flaws in the working of economic theory. People know what they want. The cost of communicating these preferences contributes nothing to turning out more goods, and so any way it can be circumvented will save resources but still permit the optimal output and distribution. As recent work in economic theory has pointed out, information is valuable, but it is not free. It may be better to spend time and effort to learn about prices, qualities, and alternatives, than to take the first trade offered or to trust others to provide you with what you want. It may be better to seek and find instead of to wait and trust that someone will provide the good you want, in the correct quantity and quality, and at the right price.

The point is, transaction costs include more than the cost of getting to the final result. They should include the result. What does it profit one to save money on a transaction but receive a good that has little value? An investment in transaction costs may yield great returns. Conversely, saving transaction costs may be very expensive. It appears that the regulations relating to water quality and automobile emissions have produced transaction costs, in inefficient use of the environment, that are far greater than the transaction costs they were designed to save. A few more savings like that and we shall need to redefine efficiency from a position of poverty. The National Water Commission (1973: 513) estimates that the cost of meeting the 1977 water quality standards, which require use of the best available technology, will be $467 billion (1972 dollars) over the period 1972–1983. Note that this significant figure just meets the 1977 standards, and only for water (not air), in 1972 dollars. Is this likely to be worth the saving in transaction costs?

Perhaps there is a broader lesson to be learned from transaction costs. Such costs are a part of the world. Nice as it would be to live without them, it may be cheaper to live with them. Consider an analogous cost—transportation costs. In international trade, the initial analysis assumes there is no cost of transportation. Iran may enjoy a comparative advantage in the production of bricks and Norway in the production of lumber. Yet trade does not take place between the two countries in these two products. Unfortunately, transportation costs keep the market from working with that utmost efficiency we prefer. Governments may order such trade or may order ships to carry the goods at a zero price to promote a more efficient economic system and correct this part of the anatomy of market failure. Fortunately, it appears that transportation costs are too tangible for such frivolity. Unfortunately, transaction costs are not. Thus, it sometimes appears that we can profitably engage in mindreading and commanding to overcome these costs; but there is no assurance, and little evidence, that the cure is better than the disease. The same holds true for the declining cost firm. Dealing with transaction costs by assuming we know desires, tastes, responses, costs, and several other variables, may be like trying to open a can by assuming a can opener.

There is another difficulty associated with regulations, although it is not normally discussed under that heading. To the extent that a person or a firm is permitted to discharge wastes into the environment without paying for that privilege, when such an action damages others, his final product will be underpriced, and inefficiency results. Regulations permit such a discharge. In fact, we shall later argue that not only is steel underpriced as a result of regulations, so are beauty and recreation. There is a more efficient solution.

If regulations are subject to these disadvantages, it is logical to seek an alternative way to deal with environmental problems. The newness of the problem, however, may explain the continued use and growth of regulations. New problems are customarily resolved in the political arena, and those concerning the environment are no exception. The decisions mean gains to some and losses to others. It is unlikely that political leaders or administrators within the bureaucracy would relinquish their right to dispense these rewards and punishments. They have not, nor does it appear they are likely to do so in the future. Each new problem seems to bring new regulations and a new agency to deal with it, such as the EPA, FEA, ERDA, EEOC, OSHA, and countless other permutations. The recent literature on the economics of bureaucracy does not predict suicide by these agencies. Regulations seem to be here to stay, notwithstanding public pronouncements to the contrary.

## Subsidies

The case for subsidies has attracted far less support, perhaps because its principal proponents have been businessmen. With a broadened base of the electorate, it has become increasingly unpopular to take action that may be construed as overtly favoring businesses over voting citizens. Businessmen have argued that by cleaning up their effluent, they are, in effect, providing residents downstream with improved water quality, and business should be paid for this service. They point out that expenditures on waste abatement are nonproductive as far as they are concerned, because these funds add nothing to the quantity or quality of the product they produce. If others benefit, others should pay. But because others are not likely to pay directly or voluntarily, businesses have asked the government to assume this burden. Approximately thirty-four states have given some form of preferred tax treatment for certain expenditures for pollution abatement (Barnett 1970: 55).

One of the most charitable of these proposals was enacted in Oklahoma in 1967. It called for a tax credit of up to twenty percent per year on pollution abatement equipment purchased by a firm, until the entire amount was written off (Barnett 1970: 64). The authors know of at least one state where a similar measure was introduced in the state legislature, and the sponsor was unaware of the difference between a tax deduction and a tax credit. A tax deduction is subtracted from the income subject to tax; a tax credit is subtracted from the tax due. Thus, an eventual one hundred percent tax credit means that the State has in effect paid for the pollution abatement equipment.

Businessmen prefer subsidies to regulations or charges, for the subsidies help meet their newly imposed costs of effluent treatment. Politicians have to weigh the benefits from the subsidy against the burdens of higher taxes to finance them. To date, however, subsidies from states to industry have been quite small, only about $120 million net from the states in 1970 (Macaulay, 1974a: 84).

Whether one supports or opposes the use of subsidies often depends on the way the basic problem is viewed and stated. If one thinks that the basic problem is pollution, it may be logical to view the firm as cleaning up pollution or producing a cleaner environment, for which it should be paid. However, if one thinks of the environment as a scarce asset wanted by many persons, then a person should be charged if he uses it, but he should not be paid if he does not. As simple as this point is, it has been misunderstood by both theoreticians and politicians. For example, we shall discuss later (p. 106) a case where Musgrave (1959: 138) proposes that people who would like to use a bridge, but find the price for its use too high, should be paid for not using it, the money to come from tolls paid by those who do. On a practical, political level, if that is the way to express it, farmers are sometimes paid for not using their land.

Firms, and individuals, may argue equally well that they should be paid for treating their garbage rather than hauling it to some dump or landfill where it would consume land sought by others. Or, that they should be paid for buying labor-saving machinery because this will release workers whose labor is sought by others. We do not pay firms for what they do not use, unless they own the asset and have sold or rented it to someone else. Rather, we require firms, and individuals, to pay for what they do use. If they choose to go to some expense to avoid using a product that would be even more expensive, that is not an economic reason for a visit from either St. Nicholas or Uncle Sam.

On the other hand, economists cite disadvantages that stem from subsidies. First and foremost, subsidies discourage business firms from finding more efficient ways to treat their wastes. If the most efficient way to treat wastes at the present time costs $100 per unit treated and a subsidy is given to meet that cost, but a new invention will permit treatment at a reduced cost of $30 per unit, the firm has little reason to adopt the more efficient process if the subsidy will be cut to reflect lower costs. Indeed, some writers have pointed out that with the right kind of subsidy, it would pay firms to go into the waste producing and waste treating business (Kamien 1966).

In addition to being a payment for wastes treated, a subsidy usually carries a right to discharge into a stream the rest of the wastes produced. This is similar to the right granted under regulations. The same criticism of regulations as promoting underpricing and overconsumption of the final product applies here, and with a vengeance. Not only can the firm use the stream and impose costs on others without having to pay for this service, but the cost of treating a portion of the wastes that results from producing the final product is also borne, at least in part, by others. There is little reason for a firm to economize in its production of waste products even though these wastes impose costs on society whether treated or discharged. Inefficiency results, and inefficiency is anathema to economists.

There is perhaps one case where a subsidy may be justified. If government regulations have been issued to require firms to treat more units of waste than could be justified by the optimum $Q_1$ in figure 3–1, firms are then required to produce a good to be enjoyed by others in society, but not on the basis of efficient use of resources. An analogous and well-known case of government-ordered subsidy is the requirement by the Interstate Commerce Commission that railroads continue to operate certain lines and trains at a loss to serve given communities. In such cases, if the government wishes to bestow a particular good or service on certain citizens, the government should pay for it. In the particular case, it should subsidize the production of water purity in the $Q–Q_1$ range in figure 3–1 (Macaulay 1974b: 1025).

*Effluent Charges*

Effluent charges constitute a third way to effect the optimal level of pollution. Understandably, they are not a favorite of business firms; it is better not to have to pay for wastes discharged than to have to pay. It is not so clear why charges are not embraced by legislators. Surely it is desirable to have more funds flowing into the public purse from which expenditures can be made. However, it must be even more profitable to lawmakers to retain their power to make benefactions in kind, through the granting of pollution (broadly defined) permits, than to face firms with new charges, which may not be an attraction to this group of voters. At any rate, with rare exceptions, legislators have not rushed to levy charges on polluters.

One popular shibboleth is that a firm should not be allowed to buy a license to pollute. Purity is too valuable to be sold in a market for mere money. This is argued with a straight face by those who daily sell for money the right to exercise command over their own lives, that is, they work for money wages, and by people who find a sprinkler system too expensive to install in homes where their children sleep. These opponents of charges confuse effluent charges with a license. When a buyer pays for each unit of a product he receives, the payment is usually called a charge. A license fee, on the other hand, is often a single payment that permits the buyer to act within broad limits. One may buy a license to fish without a limit on the amount caught, although there is usually some limit on the time for which the license is valid and the size of individual fish caught. A charge, on the other hand, is a payment for each unit received. The effluent charge is not a license without limit but a price to be paid for each unit of waste discharged.

Economists support effluent charges, however, primarily on grounds of efficiency, both in production and in consumption. If the upstream firm is charged a price of $P_1$ in figure 3-1 for every unit of waste it discharged into the stream, it will discharge only $Q_1$ units. However, the firm will be constantly on the alert to find a more economical way to treat or to avoid the creation of wastes so that it will not have to discharge into the stream and then to pay for that privilege. Economic theory and empirical evidence show that when the price of any factor of production rises, users immediately begin to find ways to economize in its use. The river is a factor of production, too, and no exception to the rule. A price will encourage economy in its use. Many options to economize are open to the producer, and price is the pressure that will encourage him to seek these out. His forebearance in use of the river will rise along with his price for use of the river.

In consumption of the final product, efficiency also rears its lovely head. Since the firm now has to pay not only for the portion of wastes it treats but also for the amount it puts into the stream at a cost to those downstream, the price of the final product will rise to a level to include all the costs imposed on society. The product will be bought only if its benefits exceed these and other production costs.

As a result of these economic advantages, economists have generally supported a system of effluent charges. This position has been cast in popular parlance as "The polluter should pay." The obvious logic of that popularization has been sufficient to add a few allies to the ranks of economists. The Sierra Club and Senators Muskie and Proxmire have spoken out for effluent charges. However, a reading of the Senators' testimony on the subject indicates they mean treatment charges instead (United States Senate 1970: 194–195). They argue that a firm that discharges wastes into a river should be made to pay a fee equal to the cost of treating these wastes at a central waste treatment plant somewhere downstream. The payment then is for treatment, not for the use of the stream with resulting damages to residents downstream. The Sierra Club supports charges, but wants them to be set high enough to discourage any use of the river for waste disposal (United States Congress 1971: 1199). Perhaps a rose is a rose is a rose is a rose, but user charges are not user charges are not user charges.

## A More General Theory of Externalities

One of the first questions that should strike the reader after reviewing the proposal for effluent charges is how this fits with Coase's analysis (1960), which stressed the bilateral nature of externalities. If externalities are truly bilateral, why are not both, or all, parties charged? If each party imposes some damage on the other and such costs should be reflected in the prices of products, why is not the externality caused by Mrs. Murphy recognized by some charge on her?

Figure 3-1 may make the point clearer. If the mill is allowed to discharge $Q_1$ of waste into the stream, it benefits from each unit of discharge as shown by the marginal benefit curve. However, the residents downstream suffer from each unit, as shown by the marginal cost curve. This point is readily seen and has led to the proposal that the upstream firm should pay a charge based on the damage its wastes cause to others. So far, so good.

The residents downstream are not enjoying perfect purity. $Q_1$ units of waste are still being put in the river. Residents downstream suffer this affliction, and the argument is frequently made that their suffering should be assuaged by compensation. But here again, the solution one chooses is a product of the way one views the problem. If people see that while less waste is being put in the river, nevertheless some is still discharged and causing damage, they may argue for compensation. Alternatively, if they note that before the upstream firm reduced its waste, $Q_2$ units were winding their way downstream to create disruption, but now there is less waste and less damage, perhaps there is room for some rejoicing. In fact, since the upstream firm has had to pay good money to dispose of these wastes in other ways and downstream residents now enjoy a cleaner world, perhaps instead of being compensated, they should pay.

In fact, all the arguments for making the polluter pay apply equally well to the residents downstream. A variation of figure 3-1 is given in figure 3-2 to help see the gains and losses more clearly. When the upstream firm discharges $Q_1$ units of waste into the stream, it benefits by an amount equal to $abc$. People downstream, however, suffer an amount equal only to $c$. If a charge of $P_1$ is levied on the firm, it will pay $bc$ but still enjoy a net gain equal to $a$. It pays because it imposes costs on others.

Consider now the downstream residents. When $Q_2$ units of waste were discharged, they suffered costs equal to $cdef$. When the discharge level is reduced to $Q_1$ and the firm is required to dispose of $Q_2 - Q_1$ units of waste elsewhere, downstream residents can now gain an amount equal to $def$ from using the stream. This comes at a cost of $d$ to the firm upstream. If one party gains from using an asset, and in so doing imposes a cost on others, he should pay for the use he enjoys. The upstream firm enjoys using the stream for waste disposal, at the expense of those downstream. The downstream residents enjoy using the stream for recreation, at the expense of those upstream.

If Coase is correct in that externalities are truly bilateral, and if it is

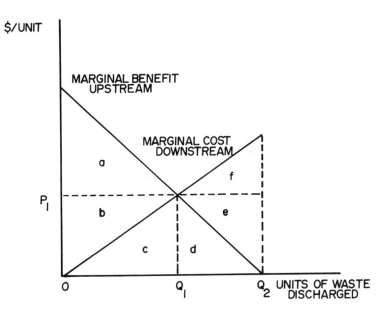

**Figure 3–2.** Gains and Losses from Environmental Use

correct that each party does indeed profit to some degree from the optimal level of pollution but that he also imposes costs on others, why has there been an almost universal acceptance by economists of the position that only one of the parties should pay? It happens, as noted in chapter 1, because of the special theory that was devised to deal with environmental pollution. Almost all people who have examined the problem have concentrated on the pollution of the stream instead of the usefulness of the stream. Traditional economic theory could easily deal with the allocation of a useful asset, but it appeared necessary to devise a new theory to deal with something that was not a useful asset—pollution. The solution was, "The polluter should pay," and this was meant the polluter in the narrowest sense of the word. It was easy to see how waste put into a stream would pollute it, but it was not possible to see how love of clean water and merely looking at it could be called pollution. Hence, only one party, the polluter (in the narrow sense of the word) paid.

The general configuration of the curves in figures 3–1 and 3–2 should be familiar to anyone who has been exposed to the first graphs that economists draw, those of demand and supply. In a way, that is what figures 3–1 and 3–2 are. But the problem is clearer if we view it in a slightly different, but familiar, context. Instead of dealing with the quality of water in a stream, we shall consider beach-front property at the end of a narrow cove. Two happy residents of the area, Humpty and Dumpty, enjoy using the beach, and each would like to build a beach house on this property. Humpty's demand for beach frontage is shown in figure 3–3 (a), and Dumpty's demand is in figure 3–3 (b). Dumpty does not want so much frontage as Humpty, but he wants most of it badly, relative to Humpty. If there were 250 feet of frontage, each could move in, take what he wanted, and live happily ever after.

Having left Eden, we sometimes find that there is not 250 feet of beach but only 150 feet. For a finite amount of some asset demanded by two parties, a presentation like that in figure 3–3 (c) was devised by economists to depict simply the problem and the solution. The problem is scarcity, which is not new. The solution is that each party gets less than he would like, and each pays a price for what he gets. Each would pay about $46 a front-foot; Humpty would get about 81 front-feet; and Dumpty would get about 69 front-feet.

If, instead, we had seen the problem as Humpty's pollution of Dumpty's use of the beach by coming in and building on it, we might have charged Humpty and given to Dumpty. But then, we would be making Dumpty the owner of the property and our diagram would consist of lines showing Humpty's demand and Dumpty's supply. That graph would closely resemble figure 3–3 (c). This approach becomes more difficult when we have three Humpties and two Dumpties (p. 107).

Still, progress toward solution moved slowly when water quality and air quality were involved. Even people who proposed a charge on only one of the parties sometimes felt that something was amiss; but this was not enough to lead them to propose a charge on each, or all, parties. Buchanan and Stubblebine (1962), and later Turvey (1963), recognized that a charge on only one of the parties would not be a stable optimum. Further bargaining between the parties would produce a new equilibrium, but the new result would not be economically efficient; so they proposed a tax-subsidy scheme. The polluter would pay a price that would produce an optimum level of pollution, and an equal payment (at the margin) to the suffering party would keep him from seeking even greater cleanliness.[6]

Baumol (1972: 309–312) later dealt with the use of charges, which he referred to as taxes, to promote efficiency; and after paying the customary homage to Coase, he proposed that the smoke-producing industry be required to pay for the pollution it created, but that the adjoining laundry be spared of charges, and denied any subsidy, since it did not pollute. In corresponding with Baumol about a policy of charges on both parties, Buchanan said that it had never occurred to him to charge both parties.[7]

The point made by Coase that both parties hurt each other needs a few refinements to give it greater precision. Although each party to the smoke problem may hurt the other, it does not necessarily have to be that way.

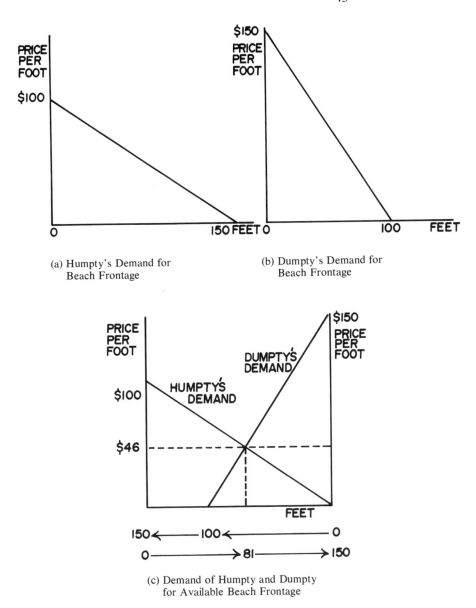

(a) Humpty's Demand for
Beach Frontage

(b) Dumpty's Demand for
Beach Frontage

(c) Demand of Humpty and Dumpty
for Available Beach Frontage

**Figure 3–3.** Demand by Two Parties for a Finite Quantity of a Good

One party may have his way entirely, so that he is not hurt at all. When
the mill has no limit on the amount of smoke it may discharge into the air,
only Mrs. Murphy is hurt, not the mill. Mrs. Murphy should pay no
charge. Similarly, when Mrs. Murphy has no limit on the amount of

freshness in the air she enjoys (that is, the mill must remove all smoke), only the mill is hurt, not Mrs. Murphy. The mill should pay no charge. However, when neither party has his way without limit, but each is constrained in some way, then each gains and each loses. Each should pay for his gain and the loss he imposes on the other.

A second refinement should stress man's answer to gravity, namely law. We have seen clearly in the past that when waste was discharged into a river upstream, gravity took it past the residents downstream, and they suffered from this waste, by either enduring, treating, or moving away from it. If, however, the people downstream put waste in the water, it could not hurt the people upstream. Only upstream people could hurt downstream people; the opposite was impossible. Then came the law. If downstream residents prevailed on the legislature to pass a law forbidding discharge of waste into the river, pollution then flowed upstream. The mill upstream would find that the river was unusable, or polluted, for the purpose it wished, just as with gravity the stream had become unusable, or polluted, for the purpose wished by downstream residents. Upstream residents now had to endure, treat, or move away with their wastes. Biologically, the law did not pollute the stream; economically, it did. If the problem under discussion is an economic one, laws should be considered conveyers of pollution as much as gravity is. Our concentration on the concept of pollution as a biological phenomenon has kept us from seeing the problem in a broader context; economists appear to have been as myopic as the rest of the population.

There is, however, an even broader base on which to build a theory of environmental use. Mohring and Boyd (1971) have characterized it as asset use. It is the basis on which the rest of traditional economic theory is founded. If there is an asset that is wanted by more than one party, it should be allocated so that the greatest good is obtained from it. Decisions on who will get the scarce food, clothing, shelter, land, labor, and capital are based on this principle. Why not the environment, as well? At least the analysis of the problem could be based on this principle.

With this approach, the quality of water in a stream can be viewed as an asset whose use is sought by two suitors. The firm upstream wants to see the quality used to transport its wastes to distant shores. The residents downstream want to use the quality to provide conditions suitable for clean living. The two uses are inconsistent, given the location of the protagonists. This is no different from the economic problem facing a landowner who is approached by two parties, one wishing to buy or rent his land to be used as a garbage dump or landfill and the other wishing to buy or rent the land to be used as a recreation site. Economic theory is admirably suited to deal with that problem. The owner accepts bids for parts or all of the tract and finally puts it to the use for which the marginal return on the last acre is highest or those uses for which the marginal returns on the last acres devoted to each use are equal. It is probable that with a sufficiently large tract to dispose of, each potential user would obtain some land for his preferred use, each would pay for the part he received, and each would suffer a little from not getting so much as he wanted—at a lower price. Charging for the use of the land promotes allocative efficiency. Yet proposals to apply these results to the environment are not usually offered by either environmentalists or economists.

In the familiar case of buying and selling land, economists raise no hue and cry that bidders for land who obtain acreage for industrial sites should compensate those who are outbid and thus forced to put their homes or recreation facilities on smaller lots. Instead, they see clearly that each user should pay for the asset he receives, and that this payment must exceed what competing bidders would pay for the same good. Two advantages derive from this arrangement. It assures us that the asset goes to the user who values it highest and therefore where it will serve society best, the usual caveats considered. Further, if users must pay for what they receive, there is a mechanism to check that benefits are greater than costs, for both will be expressed in the same terms—either money or utility.

If we view the environment as a good or an asset just like thousands of others that we deal with daily, and the parties upstream and downstream as people who want to use that asset, we have an existing system to deal with our problems. Next we shall deal with a few special problems that apply to the environment as an asset. Then we shall turn to fitting it more snugly into the existing operation of the market and the theory of property.

**Notes**

1. It is also true that the loss to those who leave should be small. If the farm were free, people would use it until the average return that each received would just exceed the earnings he could receive in his next best alternative. Under a system of charges, some would go back to this next best occupation (Worcester 1969: 877).

2. If Mudville Mill had costs as shown in table 3–2, its outlay for waste treatment equipment would drop over 75%, from an indexed cost of 500 to less than 125. And 90% of the waste would still be removed.

3. As is customary, we show this and other functions in a linear form to simplify our presentation.

4. Thomas Jukes (1973: 13) makes the same point in a different way. "One of the oldest principles in toxicology was stated by Paracelsus

almost 500 years ago: 'Everything is poisonous, yet nothing is poisonous.'" It depends on the amount of the substance in question. Even substances that are necessary for life can prove deadly if we are exposed to too much, or too little, of them.

5. A good discussion of the major issues in the administrative actions taken is provided by Hoffman (1973). The royalty and tax loss is reported in *Clean Air and Water News* (July 2, 1971: 397).

6. The case is discussed in greater detail in Chapter 4 (pages 59 ff.).

7. See page 61.

# 4 Refinements in the Theory

The theory of environmental use that has been discussed is relatively simple in concept. All parties who wish to make use of the environment should pay for the use they enjoy. A system of charges theoretically could deal with this problem just as it has been used to deal with the allocation of other private goods like motors and movies. But a few questions arise. There are ways to deal with pollution other than just treating the effluent. Can the use of charges guarantee that these other options are considered and used when they are better? What about people who move away? How shall we count their preferences and suffering from moving? In a society that stresses equal treatment of everyone, should all people enjoy equal levels of environmental quality, or should all users pay equal amounts for their use of some part of the environment, such as a stream? Or equal prices? Or different prices? The proposal calls for all users of the environment to pay, but if only the polluter pays, will that solution hold? Will it produce equilibrium? If people are charged for both waste disposal and recreation, is there not a double charge for a single good, and a resulting inefficiency? We shall see.

## Alternative Methods of Disposing of Wastes

The marginal benefit curve derived in figure 3–1 was based on the cost incurred by the mill if it treated the wastes it produced. There are other options however. The mill could have the waste hauled away and treated elsewhere; it could alter its production process to produce less waste per unit of product; it could simply reduce its output of whatever produces the waste; or it could decide that it is not so bad after all to put up with the waste at the plant instead of paying to put it in the river or to treat it; that is, purity isn't worth poverty.

Again, the problem is a familiar one. In the present case, the firm seeks a service, waste disposal, that can be supplied by different producers. Which one or which combination will it choose? In the everyday case, a firm seeking to buy transportation (cars and trucks) or transformation (machines and tools) can draw on several suppliers of the good or service it wants; and it must choose the least expensive supplier or the most efficient combination of suppliers from whom to buy. The firm buys

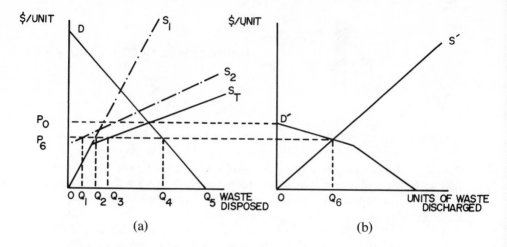

**Figure 4–1.** Choosing among Alternatives for Waste Disposed

units from its different suppliers until the marginal costs of goods from all suppliers are equal. In effect, it will sum the supply curves horizontally, determine the intersection of this total supply curve with its demand curve, and then buy from individual buyers where the equilibrium price cuts their individual supply curves. Similarly, the firm considering waste disposal will use different systems to get rid of its wastes until their marginal costs are equal. Figure 4–1 may help to explain the procedure.

Assume that the firm is faced with the problem of disposing of its wastes and has found two alternative methods. It can treat its own wastes at costs shown by $S_1$ in figure 4–1(a), or it can have the wastes hauled away and treated by a waste treatment firm at costs shown by $S_2$. The least expensive way to dispose of a given quantity of waste is shown by $S_T$, a horizontal summation of the two lines. The firm's demand for the service is a derived demand, computed in the same way one derives a demand for any other factor of production, for waste disposal is basically a factor of production. The demand line is the marginal revenue product of waste disposal.

If a stream were not available for waste disposal, the intersection of the demand and total supply functions would provide an answer to the problem. We could stop with figure 4–1(a). However, where the river exists as an option, the demand for its services is derived as a residual. Just as the domestic demand for a foreign good is that amount not supplied locally at given prices, so the demand for the stream is the amount of waste disposal services not supplied locally by other means at given prices. This is shown as the demand $D'$ for the services of the stream in figure 4–1(b). This demand is the same as the marginal benefit function in figure 3–1. The supply-like function in figure 4–1(b) is the same as the marginal cost function in figure 3–1. It measures the damage that wastes would do to downstream residents. The equilibrium in figure 4–1(b) determines the price for and the amount of waste that will be put in the stream. It also affects the price for and the amount of waste that will be treated by the different processes shown in figure 4–1(a). Figure 4–1(a) shows that $Q_1$ of waste will be hauled away and treated by a commercial firm. $Q_2$ will be treated by the firm in its waste treatment plant. $Q_4 - Q_3$, which is equal to $Q_6$ in figure 4–1(b), will be put in the stream. $Q_5 - Q_4$ is the amount of waste that will not be created because the cost of waste disposal by any means is too high to justify the production of the final goods that create that amount of waste.

The graphs in figure 4–1 show, perhaps a little more forcefully, the obvious results of a rise in the cost of using the stream for waste disposal. If residents downstream increased in number or came to love purity more (if the supply curve in figure 4–1(b) were to shift to the left), or if the government merely set a higher user charge on waste disposal, less waste would be put in the stream, but more would be treated by alternative processes, and less of the final product would be produced.

The lessons are simple and clear. A firm that is charged for putting waste into the stream will do so only if it is less expensive to dump than to treat. Dumping when it is less expensive to do so means gains for, and power to, "the people"; or, as economists put it, there is a social gain from dumping. Despite passionate prose to the contrary, society will lose less, or gain more, if it puts some $Q_6$ waste in the river and takes the money (land, labor, and capital) it would have spent treating these units of waste and devotes it to building hospitals, homes, and hula-hoop factories, or whatever people indicate they prefer by their spending habits. Furthermore, if the dumper, in order to use the river as he would like, pays a fee that is higher than other people would pay to use the river as they would like, we know that the river will serve society better by carrying away these units of waste than by looking pretty or doing more of anything else. Serving the people and doing good is the goal of economics and economists.

In this case the dirty river does more good than a clean one, just as a room that has been lived in, and looks like it, does more good than one that is spotless. Granted, we all want a room in the latter condition, but we want it so that we can live in it and convert it to the former condition. Keeping the room, or the stream, in its unsullied, readily-available state is only a modern rerun of mercantilist theory, which advised its adherents to "gather ye gold dust while ye may" and keep it in the vault. Perhaps 1977 will be a good time for a new Adam Smith.

It should be clear to the reader that the functions shown in figure 4–1 and discussed above are static representations. As prices are imposed on stream use and there is time for firms to respond to these changes and charges, the functions become more elastic. A small change in the price for using the river causes greater shifts to using treatment plants, changing production processes, and manufacturing substitute final products. However, the general shapes of the functions and the results described will continue to hold.

There is much less discussion about how the residents downstream may respond to changes in charges for their preferred use of the stream. Much of the rhetoric on the matter appears to assume an inelastic demand by those who prefer clean water. It would appear that there is no substitute for utterly pure water and air. Thus, Congress finds it appropriate to decreee no discharge into waterways, and it gives to a Federal agency the power to prohibit emission of any air pollutant that, in the judgment of the administrator, "has an adverse effect on public health or welfare" (Clean Air Act of 1970). Clearly, oxygen should be a candidate for inclusion on that list of dangerous and costly emissions because of the hundreds of millions of dollars of damage it does each year by causing rust and corrosion and because of the important role it plays in supporting fires that kill about 7,000 people a year (*Statistical Abstract, 1974*: 65).

However, just as there are substitute ways to treat, avoid, dispose of, or endure wastes, so there are substitute ways to create and enjoy beauty, provide recreation, or locate homes on beautiful sites. The analysis for downstream people is the same as that shown in figure 4–1 for upstream denizens. Only the names are changed, but not to protect the guilty— those who claim they must have clean water and air or they will die. Reactions downstream should be just like those upstream, and, more important, just the same as they are for every other good that both upstream and downstream residents buy and use. As the price of swimming in the river goes up, people find that it is cheaper to swim in the community pool, in the backyard pool, in the same river but farther upstream, in another river or lake that may be a little farther away, to substitute tennis, track, and a shower in the place of swimming, or to go to the movies.

The final solution should produce a little more of all these substitutes and a little less, or perhaps a lot less, of the previous use of the stream. If there are good substitutes for the clean water uses of the river, people will turn to them readily, and losses will not be so great. If substitutes are few for those downstream, they will outbid those upstream for the use of the stream, and the latter group will do the shifting to the substitutes. The former group will get their way for a few dollars more. Only if everyone wants to use the river, has no substitutes for its services, and must pay to

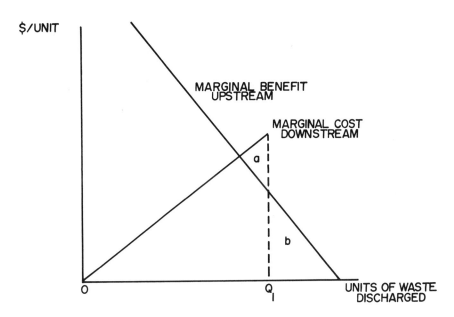

**Figure 4–2.** Marginal Effects of Stream Use When the Downstream Party Moves

enjoy his preferred use, will there likely be high costs. But then it may be better to blame not the mill, or more appropriately, people who badly want the product it produces, but Nature's niggardliness and man's intractability.

There is a subsidiary problem that has worried some writers (for example, Baumol 1972: 314). If things become so bad for the downstream party that he vacates the premises, there will be no damage from further wastes put in the stream, or from any wastes, for that matter. Should the upstream firm be charged if there is no one damaged? The marginal cost to downstream residents in figure 3–1, which is also the supply curve in figure 4–1(b), would look like the marginal cost curve in figure 4–2 under these conditions. If the demand for the use of water quality by upstream firms were like that shown in figure 4–2, the marginal cost of waste discharged beyond $Q_1$ would be zero, but there should nevertheless be a charge based on the damage the firm does.[1] The charge should be a sum at least equal to the area under the marginal cost curve. This should be a flat fee, or a charge per unit, for waste discharged up to $Q_1$, with subsequent discharges permitted at a zero cost. In computing this charge, we must beware of including all persons who would like to settle along the bank of the river. To do so would be analogous to Musgrave's proposal (1959) that tolls collected from those who used a bridge be paid to all those who would otherwise have used the bridge but were unwilling or unable to pay the toll (that is, they moved away from the bridge). For further discussion, see pp. 112 ff.

## How Much Should Users Pay?

This presents problems for both upstream and downstream users. The solution for each differs in important respects. We shall examine both groups of users to determine the conditions that should lead to efficiency and the conditions that may lead to inefficiency. First, the upstream users.

Figure 3–1 shows with some precision the amount of waste that the upstream firm should be required to treat and the amount of water quality it should be allowed to enjoy by having upstream wastes removed from the stream. It is important to note, although this importance is often overlooked or misunderstood, that figure 3–1 describes the condition at a particular point on the stream and at a particular time. Some other place or time may give different results. Perhaps the point is obvious. In locations where there are large numbers of residents downstream, the cost of discharging wastes into a stream may be relatively high, so little waste will wind up in the stream: consider the Hudson River at Yonkers. Conversely, where few people live or want to live below the discharge point, the cost is low, and waste may flow but tears will not: consider the Coosawhatchie in South Carolina.

Drawing on our constant companion for comparison—land—it is understandable that the demand and supply for land on Manhattan Island results in a higher clearing price than that for land on Mackinac Island. Further, more land is devoted to commerce and residences in Manhattan, and more to campsites and recreation on Mackinac. Land prices and forms of use differ through both space and time.

Apparently, the message has not come through for the environment. There are pleadings, arguments, and laws that set nationwide quality standards for environmental use. The argument is sometimes made on grounds of efficiency: one area should not be allowed to compete unfairly by allowing a greater degradation of its environment in the production of goods. At other times, the argument is based on equity: all areas should be treated equally so that every citizen can enjoy the same level of environmental quality.

The equal treatment that will best promote fair competition, efficiency, and maximum welfare is not an equal quality for all but an equal price for all users in a particular place at a particular time. The logic is simple.

Where a good is plentiful, its price should be low, and those who will benefit from the good should be encouraged to move in and use it for doing good. This means that in relatively unpopulated areas, where the damage from wastes discharged into the air and water may be small, firms should come in and use the air and water as the productive assets they are. It is grossly inefficient to require a firm to treat its wastes at relatively high costs when it is located in an area where the damage from wastes put into a stream is low, perhaps even if the concentration of wastes is high. Residents of the Ruhr Valley have survived, and even have profited from high concentrations of municipal and industrial wastes in stretches of their streams, while residents along the Willamette have preferred a different use.

One may ask why owners of 1975 automobiles who live in Glacier National Park or Jacksonville, Florida, should have to pay several hundred dollars each to remove ninety percent of the hydrocarbons and carbon monoxide that were produced by the 1970 models, when they live where these substances are not a problem—even though such a policy may make some sense for car owners who live in Los Angeles. Proponents of strict regulations for everyone argue that other people in distant locations will suffer, but note that biochemical oxygen demand (BOD) wastes are assimilated by streams, carbon monoxide is assimilated by microorganisms in the soil, and hydrocarbons are washed from the air by rainfall. In fact, Kneese (1966: 80) points out that if we abandoned the use of the waste assimilative capacity of the nation's streams, "[t]his might add $10 or $12 billion a year to our national costs just for disposing of household wastes."

Recall the rules for efficiency on the part of firms as they use other factors of production. The rules do not require that each firm, no matter where located, must hire only workers with Ph.D. degrees in physics, or even that a certain percent of its labor force must consist of such workers. Nor is each firm required to have land of a certain environmental quality (for example, with fifty trees per acre) if it uses any land at all. Rather, all firms in a given place and at a given time are faced with the same price for workers with Ph.D. degrees in physics, and each firm pays the same price for land that is similar and located at a given site. If some workers or some tracts of land are less costly because they have less profitable alternative uses, or if the hiring firm does not need workers or land of such high quality, the lower-quality factors of production are sought and used by the firm.

This example gives us a clue about who will argue the "unfair competition" case. The owners who find that they can be undersold are most likely to press for laws that remove the price or quality advantage of their competitors. For example, minimum wage laws remove the price advantage of certain workers who have not yet gained experience or who live in geographical areas where labor is relatively plentiful. Those who support minimum wages often plead that they do so on the basis of charitable motives. An equally defensible hypothesis is that they do so to protect themselves or their geographical area from unwanted competition. Their charity is carefully directed toward themselves. The same case can be made, and has been, with regard to the environment. Owners of land along rivers that would warrant a large payment for wastes discharged, owners of factories that are already located at these sites and would have to pay these high charges, and residents of areas where these conditions exist, will all find that old factories may have to close or move, and that new plants will tend to locate in other areas where the absorptive capacity of water or air is still high or where little damage results from discharges and emissions. Charges there would be lower. For the old area, this means a loss of jobs, a loss of value of land, and general dislocation within the community. An effective solution is to require nationwide standards for industrial discharges. This means, set national emission standards. Under these standards, every firm, regardless of where it is located, must treat its wastes to the same quality level.

Nationwide emission standards can assure residents of a region where certain activities or wastes discharged into the environment do great damage that they need not worry about economic competition from other regions where wastes do not do much damage, even if their firms are ordered to clean up or pay up. If the idea of emission standards does not succeed, however, supporters may fall back on a second choice—nationwide stream standards set at high levels. Under this policy, all streams must be kept at a prescribed level of cleanliness. Competition from clean streams is reduced since they too must meet the high standards and cannot suffer pollution. If there is somewhere a stream of such purity that it could absorb given amounts of waste and still meet national standards, people likely to be affected adversely by this competition should add another requirement: nondegradation. This means that no matter how pure the air or water is, no matter how useless such added purity is to those who receive it, and no matter how much it costs to treat wastes created, the quality of the air or water cannot be reduced.

Under either of these policies—an emissions standard or a water-quality standard with nondegradation—no single site has an advantage over another. The less costly and more efficient site loses its advantage. The Romans plowed Carthage with salt. Many modern economies plow their superior assets with laws that limit or prevent their use. Durocher's observation that "Nice guys finish last" seems to hold for laws as it does for guys. Appendix A, taken from an earlier work by our colleague, Professor James Stepp, (Hite, 1972, pp. 53–61) is a fable providing an excellent description of how politics and economics intertwine on this question.

The argument that a failure to achieve equal levels of environmental quality will create unfair competition has been extended from a national level to an international one. Some countries, primarily the developed ones, have argued for international standards for environmental quality; less-developed countries have viewed this proposal as a device to limit their development. In our own stages of early development, we worried less about the cleanliness of streams than we did about the output of worldly goods made possible by using the streams for waste disposal. Today, that feeling appears abroad in less-developed countries. One of the authors spoke with several government officials in two countries in Southeast Asia about what was being done to promote a higher level of water purity in the streams and rivers of their countries. Each responded that cleaning the rivers and streams had a very low priority. Getting food, shelter, and medical care for the people and getting some clean water into each household was much more important. Clean water going into each household and clean water flowing in the rivers are two different things, as residents along the Cuyahoga and the Hudson should know. As Eric Johnson has observed, "Pollution control may benefit fish, but it does almost nothing to improve [drinking] water for people" (*Clean Air and Water News*, March 25, 1971: 178).

The idea that equality of treatment of people requires that all persons receive equal qualitites of environment must rest on some ground other than either efficiency or logic. There may be some reason to believe that God and the law see people as created equal, but there seems to be little evidence that people have equal tastes or prefer the same goods in equal quantities or qualities. Farm boys sometimes find a return to the dairy farm filled with pleasant odors; and workers living near paper mills have long been known to comment that the odor surrounding such a plant smells to them like money. A simple observation of the differences in the quality of food, clothing, and shelter that people choose, even when their incomes are equal, should dispel the notion that everyone wants the same quality of water, air, or any other environmental feature.

It is sometimes, indeed frequently, argued that man must have air quality that permits his survival. There can be no compromise with survival. Yet the quality that permits man's survival differs greatly from the quality people talk about. After all, Los Angeles has a reputation for very low air quality at frequent intervals. But six million people lived there in 1960, and one million more people lived there by 1970.

The quality of air necessary to support human life and to permit even a reasonable expectation of old age is not the level in question. The debate revolves around having cleaner air that permits a greater range of visibility, fewer days of discomfort or illness, and perhaps improved survival rates for a small proportion of the population particularly susceptible to illnesses and diseases associated with certain environmental conditions. Even in these cases, however, if the aim is to save lives or prevent illness, it may be cheaper (meaning that we can save more lives with a given amount of effort) to worry less about the environment and more about medical care, education, or diet. The cost per life saved through environmental purity is not a well-known figure. And if the benefit is saving the lives of a few people who are particularly susceptible to certain elements, there is a question of whether this is a public responsibility, inasmuch as unlimited medical care for all forms of illness is neither a fact of life nor a potential feature of the future, neither in the United States nor in any other country in the world. May not society be better off helping the healthy than preserving the weak?

Now we turn to the question of equal treatment of the people downstream. More accurately, we turn to the question of equal treatment of certain groups who live downstream. We said previously that all users of water quality should pay the same price per unit of waste discharged or withheld at a given place for a given time. This rule applies to people whose use precludes others from enjoying their preferred use. But there is also a group who may enjoy the environment as a public good. One person can consume the good without reducing the ability of others to enjoy it; in addition, in such cases it is often difficult to exclude others from enjoying the good. Clean water enjoyed by one swimmer can be enjoyed by a limited number of others with no reduction in its value to the first person.

The theory of public goods, developed largely in the last twenty years, deals with this phenomenon. For such a good the demand curves of the individual consumers are summed vertically. Stated simply, if one person will pay $1.00 to go swimming and another will pay $2.00, the value of water for swimming at this point is $3.00. To the extent that people use the environment at one place and time as a public good, each user should pay for his preferred use according to the value he attaches to it. In this unique case, each user may pay a different price for a single good. Determining the true value that each user places on the environment is difficult, if not impossible, and has given rise to a wealth of literature dealing with the free rider.

The difference in treatment accorded users of a public good downstream from that accorded users of a private good upstream is shown in figure 4–3. If there are two firms upstream and each benefits from discharging wastes into the river, as shown by $MB_1$ and $MB_2$, the sum of the marginal benefits to society when units of waste are discharged is shown by a horizontal sum of the curves, $MB_T$. If at the same time there are two

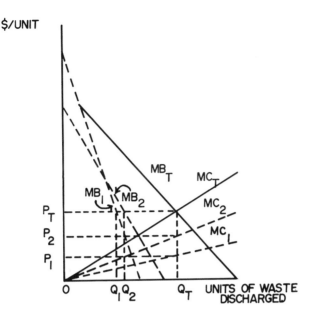

**Figure 4–3.** Marginal Conditions with Two Private-Good Users Upstream and Two Public-Good Users Downstream

residents living downstream, and each wishes to use the water quality in a
way that is noncompetitive with the other, but is still competitive with the
firms upstream, the cost to them of wastes discharged upstream may be
shown by $MC_1$ and $MC_2$. The total of these marginal costs is a vertical
summation, shown by $MC_T$. Stated in a more positive form, the marginal
benefits that the two downstream residents enjoy from their use of the
river is shown by these curves when read from right to left, and the
common usage of the water quality is shown by the vertical sum. The
optimal level of stream use occurs when $Q_T$ units of waste are discharged
and received. But, notice how the individual parties are treated.

Each of the upstream firms pays the same price per unit discharged,
$P_T$; but they discharge different amounts: $Q_1$ from Firm 1 and $Q_2$ from
Firm 2, giving a total of $Q_T$ discharged. The treatment differs for those
downstream. Each resident receives the same quantity of waste, $Q_T$, but
each pays a different price for the water quality he gets: Individual 1 pays
$P_1$ and Individual 2 pays $P_2$, which gives a total payment of $P_T$ for each
unit of waste withheld upstream.

All of this makes it seem that there is little logic or consistency in a
system of charges. People downstream who live in the same place and

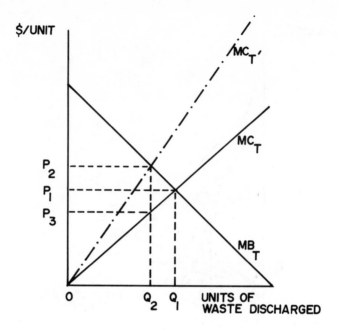

**Figure 4–4.** Stream Use with a New User Downstream

receive the same water quality may pay different prices. Meanwhile, upstream firms at a given location pay the same price per unit of waste discharged but discharge different amounts. All of this will appear odd only to those who believe equality requires that everyone gets an equal amount of a given good and pays an equal sum for these equal amounts. In the case of environmental quality, at each given point we have an equal price for all dischargers who can choose to take it or leave it. Those who choose to take more pay more. In cases where there is no choice for taking or leaving (what you see is what you get), everyone gets the same quality, and prices are adjusted to reflect desires. The result of all this, however, is the best use of the environment that is possible.

An interesting condition exists for downstream users as we have shown them. If other users move in, existing users get a higher water quality for a lower price per unit. In figure 4–4, assume that $MB_T$ and $MC_T$ curves are similar to those shown in figure 4–3. Then let another person move in downstream, so that the $MC_T'$ curve now obtains. The optimum level of waste discharged falls from $Q_1$ to $Q_2$, and the price to the older residents falls from $P_1$ to $P_3$. The total amount per unit paid is $P_2$, with the difference $P_2 - P_3$ paid by the new resident.

Notice, the effect of such an adjustment is to encourage people to congregate and enjoy a commonly-consumed cleaner environment at their point of gathering. At the same time, business firms who may wish to use the environment in a different way are encouraged to move to locations where there is less competition for their preferred environmental use. In addition to the invisible hand guiding those who prefer a common use, there appears to be an invisible foot encouraging business firms to locate where they damage least. The carrot and the stick appear once more when economic principles are used to deal with scarcity.

**The Efficiency of Charges on All Users**

Current economic thinking on the use of charges as a solution to pollution calls for payment by only the party popularly considered the polluter. If he is required to pay a fee of $P_1$ as shown in figure 3–1, he will choose to discharge only $Q_1$ units of waste, the optimum level of waste discharge. But what about those who live downstream? Even if they are not charged for their preferred use, will they not see that if they ask for cleaner water the request will increase the cost of the goods they buy much more than it will increase the value of their pleasure from the services of cleaner water? Can we not expect that they will rationally retire from the fray and allow $Q_1$ to prevail?

We shall discuss three possible reactions: first, people downstream will not remain content but will speak to those upstream; second, people downstream will not remain content but will speak to those who make the laws; and third, people downstream will not speak to anyone because they recognize the inefficiently high cost of greater cleanliness.

*Negotiations with Firms Upstream*

Buchanan and Stubblebine (1962) have dealt mathematically with the first case, and Turvey (1963) has reduced their conclusions to a simple graph. In figure 4–5, if an optimum is reached at $Q_2$ by charging a price of $P_2$ for each unit waste the polluter discharges, there is room for trading by both parties to improve the lot of each.

The firm breaks even on the last unit discharged by the firm at $E$: no gain, no loss. The price it pays to discharge the unit of waste just equals the marginal benefit the firm derives from producing that unit. If someone were to offer the firm one penny not to produce and pay to discharge this unit of waste, it would be better off not providing this last unit of output and collecting, instead, the one penny. This situation is better than pro-

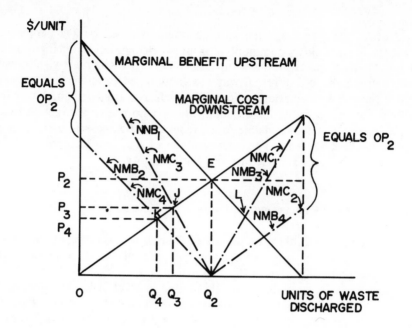

**Figure 4–5.** Environmental Use with a Charge and Bargaining

ducing the last unit and just breaking even. And because the last unit of output creates an amount of waste that damages the downstream residents in the amount $P_2$, which is presumably more than one penny, the residents are precisely the people to make an offer the upstream firm can't refuse, even voluntarily.

Turvey reaches this conclusion after deriving a net marginal benefit ($NMB$) curve for the firm. This equals the marginal benefit the firm enjoys from creating and selling the final product, less the user share it has to pay to put into the air or water the waste associated with producing the final product. Assuming that the firm is charged a price for each unit of waste discharged based on the cost that unit imposes on those downstream, the $NMB_1$ curve results. Graphically, the curve is derived by subtracting vertically the marginal cost downstream from the marginal benefit upstream. The firm will operate somewhere on this demand curve. If there is no fee or payment other than the user charge, the firm will operate where its net marginal benefit is zero—at $Q_2$. It will not stop discharging waste into the waters so long as it can reap some positive net marginal benefit.

If they are faced with no user charge, the downstream residents will seek to operate along their marginal cost curve, preferably where their marginal cost is least (that is, at a zero level of waste discharged, point $O$).

The upstream firm, however, considers that level of discharge, which is no discharge, as something like the plague. There is room for trade, based on preferences as shown by the two curves. Turvey predicts that the parties will move to a new equilibrium at point $J$. The net gain from trade, the area bounded by $Q_2JE$, will be split among the two protagonists based on the shrewdness each shows in bargaining. But while solution $J$ is an equilibrium, it is not an optimum. The units of waste between $Q_3$ and $Q_2$ create a greater benefit than a cost for society when they are put into the stream, yet they are not discharged. Charging only one party thus may lead to an inefficient solution.

If, instead of charging the upstream firm a different price for each unit of waste discharged as shown by the marginal cost curve, the firm is charged a constant fee of $P_2$ for each unit of waste discharged, the upstream firm would have a net marginal benefit of $NMB_2$, and the two parties would bargain to an equilibrium point at $K$, with $Q_4$ units of waste finally discharged. The principle is not different, but the final solution is.

Both Buchanan and Stubblebine, as well as Turvey, wrote after Coase had pointed out the bilateral nature of externalities, that each party was a cause of the problem. Yet as a solution to this problem of migration to a nonoptimum level of waste disposal when only the upstream firm is charged, they all proposed a subsidy for the parties not charged, to keep them from bargaining beyond point $E$. The idea that the downstream residents should be charged was evidently beyond comprehension. As Buchanan later wrote to Baumol, "'In my own thinking . . . I did not ever think of this sort of [double] tax at all, and it would have surely seemed bizarre to me to suggest that taxes be levied on both the the factor and laundries' " (1972: 309; Baumol's brackets). Baumol and Buchanan were dealing with air pollution, and our example involves water pollution, but the problem and the principles are unchanged. Although a subsidy can indeed produce the desired equilibrium at an optimum level, it must be based on an allocation of property rights that arbitrarily favors one party over the other. In effect, those downstream are given the river, and the fees for its use by others are collected for and delivered to downstream residents. If the subsidy is based on the cost to the upstream firm of cleaning its wastes, the downstream residents will have a net marginal cost of waste discharged shown by $NMC_1$. They will move along this curve, seeking as low a value as possible, even moving into the nether regions below the x-axis if possible. However, they encounter $NMB_1$ at $Q_2$, and trade takes place at a zero price, or, more accurately, no trade takes place. This is equilibrium at the optimum.

Turvey could have easily suggested a system of charges on, instead of payments to, the users downstream, and this would have produced net marginal benefits for those users as shown by $NMB_3$. The peculiar feature that $NMC_1 = NMB_3$ highlights an important fact that is only a restatement of Coase. The function showing the cost suffered downstream as waste is discharged into the stream also shows the benefits enjoyed downstream as waste is removed from the stream. Almost one hundred percent of the literature on environmental quality considers that the residents downstream suffer a cost, and never enjoy benefits. By the same token, one could say that when he buys or is given food, his costs of not having all the food he wants have been reduced, but he is still suffering costs. And if he chooses not to buy so much food, his costs have risen (for which he should be compensated?). One may view uncleanliness as a cost and cleanliness as a gain. They are two sides of the same coin. Smoke is bad; smoke removal is good. Food is good; food removal or a lack of plenty is bad. Economics is well equipped to deal with goods. If we have natural "bads," they can be easily converted to goods and the game continued. Indeed, goods are the name of the game, and we shall be far more successful in playing it if we think in terms of goods.

The problem with environmental economics is that it has centered around thinking "bads." The results have conformed with the approach. Since each user in figure 4–5 strives to move to that output where his net marginal benefit on the last unit consumed is zero, both parties, if faced with charges, would move to discharge level $Q_2$, where each would be in equilibrium. And this would be an optimum also. If flat charges of $P_2$ were levied on users of water quality both upstream and downstream, the curves $NMB_2$ and $NMB_4$ would result, and again equilibrium would occur at the optimum. Even if the parties could bargain further, neither would find it to his advantage to do so.

Mishan (1971: 15) raises an interesting question about all this. If the parties can bargain with each other after a charge is levied on the party upstream, why did they not do so originally and move to point $E$, with one party paying the other? He assumes that charges are levied by the government and are desirable because bargaining is unlikely to take place, and he assumes charges are levied on only the upstream polluter. Becker (1958) has reminded us that to argue that monopoly is inefficient and therefore government control is necessary is a non sequitur. The cure may be worse than the disease; and in the case Becker cites, it probably is. To argue that private bargaining is costly, so government regulation is needed, is equally unproven. Indeed, experience shows that groups are easily formed for bargaining to attain all sorts of ends that their members truly seek, from preserving the redwood forests and preventing cruelty to animals to saving the caribou from a limited sex life. There may be no assurance that private bargaining will produce a social optimum. Neither is there any assurance that a solution from charges on only the polluter will lead to a solution that is superior to one obtained from negotiating between parties or one obtained from charging all users.

## Negotiations with Those Who Make Laws and Regulations

There is another policy that may be pursued by the parties downstream when only the upstream party is charged. They may (and we predict they will) seek a quality of environment that is commensurate with the zero price they pay. They will want perfect purity. If more cleanliness costs nothing, downstream residents will want all they can use. Besides the examples previously cited showing how citizens have sought zero levels of waste discharge into air and water, there are cases in which people want all billboards along highways removed, land that is affected by strip mining to be returned to its original contour, and a prohibition on killing alligators. Each of these cases is likely to prove a corner solution, with society put in the corner, at the origin in figure 4–5.

A strange result of extending user charges to include those downstream, which means charging those who want cleanliness and purity, is that this is likely to enhance their satisfaction with the optimal solution. If those downstream are not charged, they will rationally be dissatisfied with being brought up short at point $E$ in figure 4–5, which is best for society, rather than being permitted to occupy point $O$, which is best for them at the no-charge price.

It may help us to appreciate the point if we consider what might happen if we allocated automobiles as we now allocate environmental quality to residents downstream—without a charge. In 1973, almost ten million cars were sold to American consumers. Some bought Lincolns, some Plymouths, some Vegas, and some Hondas. In most cases, owners chose a car they could afford, bargained over its price, and proudly drove their purchase from the showroom. Many who bought the low-cost cars were as happy and proud of their new wheels as those who came out with luxury liners. People got what they could afford and went away reasonably happy. But if the government had decided to allocate all cars to Americans, at a bargain price of zero dollars per car, what might have happened? If the government had shown unbounded omniscience by allocating automobiles precisely as the market would have done, it is possible that some would have been happy. It is even more likely that nearly every recipient would have considered his own sterling qualitites, his own worthiness, and his obvious need of a car commensurate with his caliber; and he would have concluded that in the great give-away, he would get one of the better vehicles—surely several steps above his ordinary neighbors. People who received Vegas, even though that is the car they voluntarily would have bought, would assume that they deserved at least a Pontiac. All up and down the line, Santa Claus would bring not Christmas cheer but consumer misery because of The Big Rip Off, or "Deals on Wheels" as it probably would be called. As a companion for the National Welfare Rights Organization, a National Runaround Rights Organization would develop. The frustration factor would turn even an optimum level of purity into a nonoptimum level of social welfare.

## No Further Action Taken by Those Not Charged

If people downstream were not charged for the water quality they received and if they made no overt effort to alter decisions on allocating water quality, could we expect any departure from $Q_2$ in figure 4–5? Probably we could. Two forces would operate. First, the fact that downstream users did not have to pay for the good they received, clean water, would give them more money to buy more of the goods they want. This income effect would surely be evident in a greater demand for the good they preferred to begin with. It might also show up in a greater demand for the goods that create a demand for waste disposal. The predicted effect, therefore, is uncertain, but probably there would be a shift upward in the *MC* curve in figure 4–5 as the richer people felt they were hurt more by wastes discharged or would be willing to pay more for their preferred use.

A second effect might stem from the purchase by downstream residents of goods and services that complement the free water quality they receive. If they do not have to pay for clean water, they can afford to spend more on boats, swimming gear, picnic supplies, and related goods. When they have these on hand, the value of another unit of purity rises over what it would have been without these goods.

To try this idea in a different context, suppose business firms had to pay for the gasoline they use, but private individuals who were owners of passenger cars did not. Then suppose you were given the task of setting the price of gasoline to trucks and the like so that social welfare would be maximized. Would the result differ from what would have happened if every user of gasoline had to pay? With private transportation reduced in cost, more cars would be bought, and the value of gasoline to passenger car owners would rise.

Finally, we turn to a general question of whether downstream residents would settle for $Q_2$ of waste discharge in figure 4–5 because they know that lower levels will result in higher charges for the final product produced by the upstream firm, and these charges are more than the value of purity. Experience does not justify great optimism. Consider what people did when the question was phrased in reverse. Previously, as more waste was put into the stream without the payment of any fee, the final product could be sold more cheaply. But the people downstream paid higher and higher costs because of greater levels of waste discharge.

Would not buyers of products realize that more and cheaper purchases of goods meant more waste, and that the cost of the waste was not worth the saving in the price of the product? Perhaps, but the people who support a zero charge for downstream users are the same ones who argue that upstream users and their customers overconsumed the stream when they faced a zero price for stream use. Consumers did not stop buying the cheap product, even though it meant even higher environmental costs.

Actually, there is little reason to expect most consumers of a product, say textiles, voluntarily to forego consuming the product in order to raise environmental quality for some users. This is true because the two groups are normally not the same. The externalities are not internalized. Consumers of textiles live in Michigan. Those who produce the good and suffer the dirty water from its production live in the Carolinas. Carolinians could forgo consuming any cloth, but this would have little effect on their streams. People who live in Michigan have no reason to reduce their consumption of textiles. Hence, few people find it in their own interest to forgo consuming textiles, even though the environmental cost of the last units is high. Similarly, producers of automobiles, production of which may also cause dirty water, live in Michigan, while most consumers live elsewhere.

Could the problems be solved by having national standards so that the cost of clean water desired by residents of both states would be readily recognized in higher prices of all goods produced? Perhaps, but the link is feeble, and it may not be recognized readily. Translate the problem from water quality to labor. Suppose that every household in the United States could have household labor at a zero cost, but that every household would receive the same amount of this free labor. With a finite supply of labor, as more labor was used in households, less would be used by business firms and the goods produced would have higher prices. To what extent would households limit their demand for the free service, recognizing that more free dusting meant more expensive Plymouth Dusters?

When the group affected both beneficially and adversely by waste discharges is small, the effects are more likely to count. Thus, chicken farmers near Los Angeles apparently overcame the objections of neighbors to the unpleasant odor of chicken farming by selling eggs to the neighbors at reduced prices. When sufferers and beneficiaries are not the same individual or members of a small group subject to personal social pressures, each person tends to seek his own self-interest regardless of the cost imposed on others. Prices, and only prices, convert this self-interest into society's interest. Thus, all users of the environment should pay a sum based on the marginal benefit they enjoy from their preferred use.

**Note**

1. In the particular case shown in figure 4-2, there is a range in which the marginal cost of the units of waste discharged exceeds the marginal benefit they create. However, after downstream residents move, marginal benefits exceed marginal costs. Only if the excess of benefits over costs (area *b*) exceeds the earlier excess of costs over benefits (area *a*) should waste be discharged beyond the original intersection of the two curves.

# 5 Rights to the Environment

## Property Rights and Externalities

So far we have described a market solution to the problem of environmental scarcity. The focus is *use*, how we can maximize the benefits that come from using our scarce resources—all of them. There is a direct and fundamental property rights analogy applicable to our discussion. With the evolution of price, there must be an assignment of property rights. The strength of the market is limited to the strength of the property rights system.

Starting with the textile firm discharging its waste into a remote river, we described *common access property*.[1] No one person or group of people owned the rights to use the river. As we described that first situation, there was no reason for anyone to spend time or money to obtain those rights. The effort would have been useless; the quality of the river was not scarce. Both the city in the far reaches of the river and the textile mill used the resource as if it were a gift from nature. Some enjoyed dumping, others enjoyed drinking and viewing. The river was free.

As the city grew and apartments developed, frustration entered the picture. The textile-mill operators became concerned about the apartment owners. "Will they scream about the sizing in the river?" The apartment dwellers began to sense the presence of the mill. "Does your water taste funny?" As the resource became more congested, the externalities became more pronounced. But during all of this, the price of the right to use the river was zero.

Eventually the conflict became so intense that something had to be done. Government entered the picture. The common property was on its way to becoming *public property*. Predictably, a Clean Water Council could develop and begin to express concern about the abused river. With concomitant legislation, the river's use would move into the spectrum of property rights. Then it would become public property. Discharge permits, environmental impact statements, and quality standards would be defined. Use of the river would be regulated. But the explicit price of use would still be zero.

Under public ownership, it is likely that frustration would expand, even if the regulatory authority made its allocation of the water quality at the exact intersection of the marginal benefit and marginal cost curves. In the absence of prices, people would want to use the stream until its marginal benefit equalled the price charged. Frustration would double. Public property rights could solve the allocation problem initially, but not finally.

We have suggested that all users of scarce environmental rights pay for them. To do this would require another transition in the scheme of property rights. The rights to quality would become *private property*. Only then would a market for the scarce resource develop. Price could then allocate the rights and remove completely the frustration of the competing parties.

Our analysis has served to point out the forces that give rise to property rights. First comes crowding. Then follows frustration. Finally, there is demand. The response from the property rights side is crucial. First there was common property, then there could be public property. Finally, there may be private property.

Although the analytics of resource scarcity and property rights evolution are relatively straightforward, the real transition is not. The development of priced property rights to land is still not complete. Even though centuries have passed, much land is still held in the public domain and many uses of private land are regulated or publicly owned. A specific example is evident in the state of South Carolina, where some 11,000 acres of coastal wetland are publicly owned (Hixson 1975). The ownership goes back to the seventeenth century and grants by the King of England to the original Lords-Proprietors who settled the colony. Today this valuable resource is leased to fishermen for the administrative fee of $15.00 per year per acre. There is no supposition that the market value of the land is so low. But it is possible that the restrictions tied to use of the land dictate a negligible or even negative price. In any event, economic value is undetermined.

Thousands of acres of federally owned land are allocated by various administrative agencies. Sometimes, particular rights are identified and sold on a competitive basis. Thus, drilling rights for oil, grazing rights, and timber-cutting rights are subject to a market mechanism. The evolution described above can be observed at work; it is not complete.

In terms of stages of development, we have suggested the property rights process as beginning with common access resources, moving to public ownership, and finally changing to private ownership of rights.

The movement between the last two stages can be broken down further. During the latter part of the second stage, certain scarce characteristics of the resource in question can be identified. Rights to these characteristic uses may be leased, rented, or sold on the basis of some predetermined total amount. Usually, these rights are not transferable. Thus, control over the total impact or use is maintained by the public owner. Still, individuals can make efficient moves toward an equilibrium position.

**Table 5–1**
**Stages of Property Rights**

| Type of Ownership | Characteristics |
|---|---|
| I Common Property | Common access. Individuals and firms use resource without paying. No interdependencies recognized. No crowding. Transition begins when crowding is first recognized. |
| II–A Public Property | Scarcity recognized. Rights to property identified by government. Allocation determined by administrative action. Laws, regulations, licenses used to allocate. Transition to III begins when cost of administration exceeds costs associated with private rights identification. Price is not used in allocative process. |
| II–B Quasi-private Property | Public decision makers identify certain rights to use. Amount of total rights to be allocated determined by authority. Rights are then rented or sold with restrictions on use. Finally, rights when issued may be sold to other users. Price plays some role in allocative process. |
| III Private Property | Individuals purchase rights from other individuals. Property rights are enforced so long as they have value greater than enforcement cost. Laws may dictate broad restrictions on use, but specific rights are not restricted as to use and resale. Price plays vital role in allocation process. |

Another step between these stages occurs when the predetermined rights are sold and made transferrable. Individual owners can then hold rights as an investment and thereby recover their earnings, if necessary, by making later sales. Still, full use of the resource is controlled by those who determine the amount to be marketed.

The final stage is reached when all rights to the resource are sold without restriction. There is no further ownership exercised by the public. One is unlikely to find an example where every conceivable right to the use of a resource has been sold without restriction. There are degrees, however, and this is the question to consider. To what degree will the market be used in allocating scarce resources?

Table 5–1 summarizes the stages of property evolution. Though the table is simplified, we shall see many situations that fit its model.

**Environmental Resources**

Turning now to that specific endowment of goods termed "environmental," we can observe property rights in evolution.[2] As mentioned above, the process is certainly not complete. In some cases, resources have just moved from stage one to stage two. Other cases have reached the latter half of stage two. And in some instances, we shall see evidence of the early development of stage three. The process takes time—decades, centuries, or longer. But it does occur.

Perhaps the best documented case where property rights have been identified and used to allocate water quality is the Ruhr basin of West Germany.[3] There are eight large associations dating from the early 1900s that are involved deeply in the management of water quality. Prior to the establishment of the associations, or *Genossenschaften*, water quality in the heavily industrialized Ruhr was the result of city, firm, and individual behavior. Water quality was a common access resource. Those using it placed a value on it, but at the margin there was little that people would do for improved water quality. It wasn't worth the time or cost. Eventually, the users of the common access resource became critically aware of one another. A typhoid epidemic resulted from bad water, and public-utility-like associations were formed and charged with managing the newly recognized scarcity. Stage two began. The scarce property rights to water quality were now appropriated by governmental bodies.

The approach used today for operating the association comes quite close to stage three. Membership in the associations is required of all waterworks and all firms that take directly or indirectly more than 30,000 cubic meters of water annually from the Ruhr and its tributaries. Voting rights are distributed to the members in direct proportion to the contributions made to the expense of the association. Payment is required from all who use water quality by discharging waste and from those who withdraw water for drinking. In this way, the Ruhr associations come close to the optimal system for charging all users for the costs they impose on others.

Since the scarce water quality is recognized, limits have been set as to how much quality will be used in this manner. All users who discharge wastes pay to defray the costs of operating common treatment plants, or, if they choose, individual plants can treat their own waste and escape certain charges. In many instances, charges are based on factors other than the amount of waste discharged. They are not uniform.

To gain control of water quality, the various associations must have some control over land use in their respective regions. Charges for discharging effluent into the streams tend to help. If water quality is relatively scarce in one location, a high charge will reflect the scarcity. All costs considered, firms will tend to locate where profits can be maximized. The frustration associated with zoned land use or some other regulatory system is minimized. Individuals react to price in a predictable fashion.

The administrators of the Ruhr association are convinced that economic incentives, prices, are powerful in their effect on the behavior of these water quality users. They are seeking ways to bring recreational users into the association. In other words, they would like to remove the view that water quality in the Ruhr is common property.

It is not surprising that other innovative uses of price are found in the Ruhr.[4] Additional reservoirs are often built to manage water systems. The Ruhr is no exception. Of course, reservoirs bring benefits and costs to the receiving location. Recreational opportunities may be improved. At the same time, valuable land with improvements may be covered by water. There is a location problem. This, too, is a property rights problem, if there is more than one interested party. Who will obtain the newly defined property? How will it be allocated?

Historically, the location of government facilities has been determined by a board or commission. Usually, opinions are obtained, hearings held, and petitions read, but ultimately, an administrative decision is made. Under the circumstances, there will likely be substantial frustration. There have been instances, however, in which bidding was used. For example, the capital of the State of Georgia was located in Atlanta by a process that included bidding. Various cities had the opportunity to offer a combination of money, land, and facilities for the coveted political capital. The City of Atlanta outbid the rest.

In the Ruhr, a similar problem was settled by selecting several potential sites for a reservoir; then the receiving areas were allowed to bid for the right to have the facility in their area.[5] The price system enabled all interested parties to express their interest in terms that could be understood and compared. The region bidding the most or charging the least received the facility.

It must be recognized that the Ruhr system is a partial approach to a full market solution with private property rights. The right to discharge in a given stream must be purchased, but it cannot be resold by the owner.[6] It is a lease or rent, not ownership. Additionally, the quantity of rights made available does not necessarily coincide with a point of full equilibrium. The quality to be attained is determined by the authority, not the market.[7] However, there is an element of competition between the eight associations. Some set high limits on acceptable waste; others set lower limits. Given the competition and ability to substitute locations, it is possible to maximize efficiency rather than frustration. The use of prices and property rights assists in communicating happiness. Water quality is viewed as something of great value in West Germany. The price system assures that its value is known.[8]

## Air Rights

Scarce environmental resources include many things. Even air space itself has become crowded. Some people may say that their views have become polluted. Others may exclaim, "What an impressive skyline." The problem is the same: a common access resource is used to the point where the last unit brings little or no satisfaction or productive benefit to one user but a large reduction in benefits to another user. We can imagine lessons from history that might have involved the obstruction of wind passages for windmills. There have been cases in which tall structures cast shadows on sun lovers. Someone was damaged in the process. Surely the day will soon arrive when rights to solar energy will carry with them the right to unobstructed solar radiation.

One may even suggest that all zoning regulations are in fact regulations of air space. Once a structure is above ground level, its appearance gives a first indication of land use. This would be stretching things unnecessarily. Air rights are a special case.

The development of property rights to air space has almost reached stage three in certain zones of New York City.[9] Much like the Ruhr system of water quality, the City of New York has determined the average vertical density to be allowed in certain districts. Thus, horizontal and vertical densities are regulated. However, air rights to vertical density are owned and can be sold, transferred to another location within a given district. The owner has reason to conserve the use of air space, for it has economic value to him, value that can be liquidated by choice.

The New York approach is interesting in that the transfer system has allowed certain lower historic buildings to remain in high density areas. Without the ability to transfer the scarce air rights to other locations where all rights were allocated, the owners of certain property would have had their property reduced in value by density zoning, or reduced the original property to rubble in order to build high-rise buildings. In some cases, banks have purchased air density rights and held them as an asset or offered them for sale to other parties. A market has emerged.

Just a slight use of the market brings the conservation of scarce resources. In the New York case, people who preferred old buildings have not been penalized completely because of their preferences. They have been allowed to purchase their desired use of a valuable resource. In the same way, investors who profit by building tall structures have a market through which air rights can be purchased.

## Environmental Property Rights and Land

As with water quality and air space, property rights to land use are in an evolutionary process. The zoning process is evidence of this. Calling for a theory of zoning, Dan Tarlock (1972: 17) had this to say about the property system used with land:

Contemporary zoning should be conceptualized as a system of joint ownership between the public entity and the regulated. It is a form of joint ownership in which the owner of the fee retains possession and the right to manage subject to veto by a co-manager, the public entity.

The statement fits stage two of our property rights classification system. Again, we see the latter part of stage two. In fact, stage three is in sight and has been reached in at least one instance.[10]

Instead of discussing joint ownership, we prefer to continue our analysis by dealing with characteristics, bundles of property rights, associated with a physical resource. In the case of land, the potential bundle of rights is quite large. Of course, the bundle grows: that is, new uses to land become recognized with technological change. Each new use becomes a new question. Will it be marketed as a private property right without restriction or will it be appropriated and limited by the public?

The answer to the question depends on relative scarcity or, put differently, the effect that a particular use has on other parties, parties with no voice or influence in the decision. At first it is a problem of common access.

Some specific examples may help to illustrate the point. If landowners are sparsely distributed in space, many untested rights are available to them. They can discharge waste, solid and liquid, on the property; they can build structures of varying shape and character; they can burn rubbish, generate noises, dig holes, stir up dust. In fact, every potential use of land may be developed. Any particular person having all the land he could possibly use will seek to maximize the use of his resources. All uses will be exploited to the point where the marginal benefit of each becomes zero.

With a growing concentration of people, greater densities on land, the potential rights associated with land use are questioned. One landowner becomes affected by his neighbor. The common access use becomes crowded. It becomes physically impossible for several owners to enjoy the potential rights of their property. The law of nuisance crops up. Rights are clarified. Restrictions, both legal and economic, begin to allocate certain rights.

In such a setting, the legal standing of zoning was affirmed. The rationale used by the Supreme Court in *Village of Euclid v. Ambler Realty Co.* turned on the prevention of nuisances prior to their occurrence.[11] Those activities called nuisances were in fact a result of crowding, or conflicting use of a common access resource. The absence of private property rights, owned with certainty by particular land owners, created a frustration called a "nuisance." If the rights had been identified and made a part of the other rights held privately by landowners, private negotiations, carried out voluntarily, could have settled the question. Perhaps the cost of doing this was thought excessive. Zoning—public ownership of the scarce right—was chosen as the solution.

For the most part, allocation of scarce zoning rights takes place on a zero price basis. One cannot legally purchase a particular zoning classification, nor can one purchase release from rezoning.[12] That is, if an owner of land zoned as residential tries to buy the rights to use his land commer-

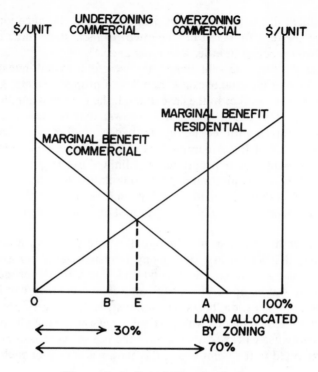

**Figure 5-1.** Land Use Analysis

cially, he cannot. Ths political process—hearings, petition, court cases—is the avenue open to him. If a party wants to keep his land and the rest in his neighborhood under a particular zoning classification, he must use the same political process.

To illustrate this, we can use our now familiar diagram. Figure 5-1 shows one curve labeled marginal benefit from commercial use and one for marginal benefit from residential use. The horizontal axis measures land available for any use. We could think of the quantity in absolute or percentage terms. One hundred percent of the land in the jurisdiction analyzed could be zoned commercial, or the reverse could be true. In the absence of a charge and without legal restrictions, parties seeking commercial use would like to move to the point of zero marginal benefit. The same is true for residential users.

Once again, we see the familiar equilibrium point. Land use can be divided by a zoning ordinance on the basis of the intersection. In terms of the diagram, forty percent of the land would be commercial and sixty percent, residential. Such a fortuitous division would be economically sound but, in the absence of charges, would probably not hold. Recipients of the valued rights would still want more rights. If price remained zero, they would demand more at such a low price.

There are two vertical lines in figure 5-1. These show the economic effects of a zoning allocation failure. Overzoning or underzoning must be expressed in terms of one use. We have chosen to express this in terms of commercial use. Too much commercial land implies that the last few acres put to commercial uses would have served the community better if they had been designated for residential use. The diagram shows this. At allocation $A$, residential demanders could outbid commercial users for the scarce zoning rights. The reverse is true at point $B$. However, in the absence of prices any zoning decision will create frustration. Equilibrium cannot result for the interested parties unless they pay the price indirectly by purchasing legal services or by using their own time in obtaining a favorable decision. These costs reflect accurately the value of marginal rights gained and lost.

There is an alternative to the public ownership and regulation of certain rights to land use. Like the scarce air rights in New York or the water quality rights in the Ruhr, the rights to land use can be determined. Some parameter can be identified; for example, the average density of land use can be established. Then, shares in this density could be set and assigned to all landowners in a community or sold to the highest bidder. If resale of shares were allowed, the market would allocate the rights. Individual owners could hold their rights or sell them. Use would depend upon the economic cost of holding the scarce rights. The price of the rights would rise and fall with the market conditions.

This sale of rights is under consideration by land-use planners today. Development rights are discussed as an alternative or appendage to the zoning process.[13] It is a movement toward stage three in property rights.

Another development in land-use controls calls for market forces to handle the problem of rezoning or initial zoning. Compensatory zoning is the suggested approach for avoiding the taking of property issue (Bosselman 1969). The procedure is simple. In this scheme, an owner of land is assumed to have the right to use his land as it is presently zoned. That is, his zoning right is certain. If the property is commercially zoned, then the owner should expect a reward for holding land for that use. To deny that use is to take away a property right without compensation. Compensatory zoning establishes a process whereby any prior zoning classification has to be purchased. Under this option, planners will be very careful in redefining land-use patterns. The changes have often been costly to the landowner; now they will be costly for those who make the changes.

Compensatory zoning is a recognition of property rights, but it is rather restricted. Only the public authority can buy. Thus, individuals who see some value in rezoning certain property must bring pressure indirectly. The process is one-sided, but it is a move in the direction of marketable rights.[14]

Privately developed communities probably carry us as close as we can get to a complete market solution (stage three) to environmental rights and land. Cities such as Columbia, Maryland, and Reston, Virginia, are actually firms that have taken into account the interdependencies between land parcels and have acted to maximize profits on the basis of consumer demand. An interesting aspect of these communities with totally planned environments is the fact that the rights were purchased, not zoned or regulated away from private owners. The managers of these city-firms act much as any multiproduct producer acts. The correct mix of activities, land use, and aesthetic qualities is carefully sought. Individuals who see what they like buy the product. Those who prefer some other package of property rights go elsewhere.[15]

James W. Rouse, the original developer of Columbia, Maryland, summarized the idea in a presentation (1973):

It is through the market place that a free people can best make the complex judgments of how, where, and when they wish to spend their earnings. A continuing examination of profitability is simply a responsible attempt to perceive the market place votes and respond to them.... It hauls dreams into focus with reality and leads to bone and muscle solution. It gives integrity to the ultimate plan.

## Natural Resources and the Market

If one examines the property rights structure associated with natural resources, one finds an element of public control in each case. The degree of public control seems large relative to other goods in an economic system. It is more than just the limit separating criminal activity from some broad category of legal behavior. The system of property rights itself is unsettled. Think of a system of variable rights, constantly in a state of flux with many exceptions, many special cases, much leeway for arbitrary administrative decision-making.

As we have shown in this chapter, land typically has zoning, air rights are allocated by commissions, water is regulated by authorities, and other amenities are captured in some control network. The cases described here are the most advanced in their use of property rights, but there is still a large degree of control.

Is there something different about natural resources, something that seems to preclude a highly developed use of property rights? Perhaps the difference relates to the first stage of production. There is a difference, if we go back far enough.

In the discussion of property rights by such philosophers as Locke, Rousseau, and Bentham, we find a relationship established that links the labor of man to his property. The so-called doctrine of natural rights implies that man has a right to his production, that our product is an extension of ourselves. The production process itself produces property rights.

Neoclassical economic theory developed a framework that explained a basis upon which production would be distributed. In terms of property rights, we may paraphrase and say that the owner of capital received certain property rights related to the marginal productivity of capital, labor received a set of rights to output on a similar basis, the owner of land received rent, and the entrepreneur received profits. A distribution of property rights, rights with certainty, springs from production. Perhaps this production-linked process settles many questions about ownership; even the rights to production or economic profits are determined in the process. But there is less certainty about environmental resources.

The goods we term "God-given" are not produced in a strict sense by owners of labor, capital, and land. They are an original endowment. Discovery and appropriation have determined ownership patterns, but these seem to be less acceptable than production-determined rights.[16]

Because the supply of environmental resources is neither price-elastic nor effort-elastic (that is, the resources exist and are not easily reproduced), it is possible to move ownership with greater ease. The example of oil reserves comes to mind. Spokesmen in various countries refer to "our stock" of petroleum as though it were the property of the public. How unusual it would be to hear references to our stock of jar lids, or automobiles, or barbed wire. We hear similar references to our environment, our wild rivers, our air and our community. Patterns of ownership, property rights, were not clear when these things were produced, so unsettled rights continue.

In another sense, in our discussion of property rights, we have observed a process that has taken long periods of time. There must have been a period when all resources were common property. All goods, including food, clothing, and shelter, were gathered from nature. At first, everything was "God-given." Nothing was produced in the modern sense of the word. Eventually, however, scarcity dictated production. Food was harvested after fields had been cultivated. Property rights became more certain because of the production effort involved. Stage three arrived centuries ago for many products.

This may explain the problem faced in dealing with environmental quality. We have realized just recently that we cannot go out and pick from nature what we need. It must be rationed. Once we begin to produce environmental quality, property rights will become much more important and more certain. People recognizing this will demand that prices be charged for the product.

It may be impossible to hasten the progress of property rights development. That, too, is a market phenomenon. A solution to environmental questions will develop when ownership patterns are clear, when river basins are operated by profit maximizers who must cover their costs in managing water quality, when land users must pay the owner of zoning rights before using a scarce resource, when "recreationists" receive a bill for using the scarce product of an owner of the environment. Once prices appear for environmental rights, society will know that these resources are valuable. They will be treated accordingly.

**Notes**

1. We have chosen to identify stage two as public property and then to go further in describing transitional stages that seem to link the second and third stages of property. Dales (1972) has taken another approach in a somewhat less structured analysis.

2. Robert Haveman (1973) has suggested that there is a fundamental difference between the common-access property question, congestion, and pollution. In our opinion, from the standpoint of economic use and efficiency, the three are synonymous. The difference described by Haveman used differing assumptions about alternative locations for individuals facing crowding and pollution. Competitive alternatives cause the three situations to converge.

3. The basic source for discussion of the Ruhr system is Kneese (1964: 160–187). However, further elaboration is found in a thesis by Joyce Eastman (1973) that included material obtained from the Ruhr association. An extensive review of literature on the subject is included in the report by Barnett, Shannon, and Yandle (1974).

4. The following discussion is based on conversations with the managing engineers of the Ruhrverband when one of the authors visited the Essen headquarters in 1973.

5. For a discussion of a similar market approach to another public decision problem, see Yandle (1970).

6. The system of effluent charges used in the Ruhr has been described as a tax or surcharge by some writers and as a treatment fee by others. There are implicit property rights under any of these names. However, the more clearly property rights are identified and the more certain their enforcement, the greater will be the resulting economic efficiency. Because of the somewhat static nature of the Ruhr system of charges, Ferrar and Whinston (1972) have suggested an amended approach. It deserves consideration. A discussion of this piece by McFarland (1972) offers other pragmatic ideas.

7. A proposal dealing with this specific problem has been developed by B. W. Mar (1971).

8. The view that prices may be useful in achieving an efficient use of the environment has not been popular. Although the idea is almost fifty years old, support has only reached the point where legislation just may be considered in the United States. A recent report by the Committee for Economic Development (1974) recommends that economic incentives be used. This organization represents a large cross section of viewpoints, so the recommendation may reflect a new consensus. However, the report contains a mistaken concept of charges. As it says, "The major thrust of our report is to bring environmental considerations within the market framework, insofar as is possible, through what has been called the 'polluter-pays principle' " (p. iii). The confusion revolves around property rights. If the rights are scarce, all users must pay. Effluent charges are not penalties for pollution any more than the price paid for a loaf of bread is a penalty for polluting another person's nutritional environment.

9. For a discussion of the New York plan, see Jonathan Barnett (1974).

10. The highly documented case of Houston, Texas, and its use of covenants between landowners instead of zoning appears to be the major exception (Siegan 1970).

11. For a discussion of this landmark case, see Babcock (1966: 3–6).

12. While the market price may be zero, this does not mean there is no cost associated with getting scarce zoning rights. As Marion Clawson observes, "[L]and developers who benefit from rezoning do not wait patiently for the plums to fall into their open mouths but vigorously shake the tree." (1971: 5).

13. For discussions of development rights transfers, see the January 1975 issue of *Urban Land*, which was devoted to the subject. A specific discussion of development rights in an environmental setting is found in Costonis (1974).

14. This idea is treated extensively by William Whyte (1968). For example, in the introduction the author states: "The increased competition for land use is not a force for blight; it is a discipline for enforcing a much more economic use of land, and a more amenable one" (p. 10).

15. For discussion of the new city-firms see, Gurney Breckenfeld (1971) and Shirley Weiss (1969).

16. The unsettled position for natural resources in economic theory may be illustrated by controversies that have raged through time. For example, Henry George stated that all citizens had "equal rights to the elements which Nature has provided for the sustaining of life—to air, to water, and to land" (1948: 36). Of course, his solution turned on establish-

ing property rights, rights that would be appropriated by government and sold to the highest bidder. The revenue obtained would be used to finance governmental services. Alfred Marshall had similar thoughts about land and .expressed the idea that "all land shall become the property of the State after a certain time—say 100 years hence.... At the end of that time the State ... might again sell the usufruct with any new conditions on its use that might then be desirable in the public interest" (1969: 205). For further discussions of these ideas and others, see Yandle and Barnett (1974).

# 6

## Markets and the Future

### Efficient Markets and Preparing for the Future

What about the future? Wise men have always been interested in the future, and the gift of prophesy has been sought throughout recorded history. The birth of science, beginning with astronomy and its associated mathematics, followed important mystical studies of the stars—astrology (Zolar 1972). The focal point was the future, and the goal was certainty, the assurance that life tomorrow would at least exist and, possibly, would be better. If only we could assure ourselves of good fortune in the decades to come, perhaps we could address the problems of the present. The reverse may be true: the problems of the present will determine to a large degree the conditions of the future.

Concern about man's environment includes considerable concern for the future. Will our children or great-grandchildren have access to the commons as we have? If they do, will they value it the same, less, or more than we? Should we save continuously for future generations and caution others to do the same? Or would some final greedy generation come along and enjoy rather than save for their future generations? The game could go on forever.

Certainly, some common-use resources will become more crowded, or less plentiful and even more crowded, in the future. History describes a pageant of scarcity. There is no reason to believe in a reversal of either population growth or income growth, those driving forces that call for more use of scarce resources. Surely more and more common-access property will be allocated to public or private ownership. The pressure to conserve, to be efficient, will be greater, not less; but so will the pressure to use. Then how can happiness best be assured?

There are two distinct problems to think about. First, to the extent that present users of environmental goods impose costs on others in areas where no market exists to reflect demands, probably there will be too much of one activity, too little of another. This has been the theme of chapters 3 and 4. For example, the operators of strip mines may be able to turn the earth's surface to obtain coal and ore and, in so doing, allow overburden and tailings to wash onto farm land. In the absence of a market to handle this conflict, there could be too much copper, not enough corn. The future would be distorted by inefficient present use.[1]

However, as pointed out earlier, present decision-makers can be forced to consider the costs they impose by making them pay. Suppose we have such a system. The spillover problem is solved. Now, what about the future? Shall we run out of ore because of nearsighted, profit-maximizing, present-money lovers? Or shall we underutilize the ore because of farsighted, future-money lovers? The motivation seems the same. Only the hats are different. Some look black; others look white.

We must realize that however we approach the future, only those present now will be able to point the way. If we leave the job to one person, it may be the one who asks, "What have future generations ever done for me?" That sounds callous. Or it may be one who says, "All our natural resources must be saved for our great-grandchildren." Can we balance these views?

If property rights are assigned to the resource in question, efficient markets will take into account many decision makers, many hats of many colors. They will, in turn, take the future into account. Of course, this happens every day; but remember, it's not perfect. However, the same degree of care will apply to future values as to present ones. There is a significant difference between the two. Greater uncertainty prevails when decisions about the future are made. Unless a new Nostradamus appears, we shall have to make the best projections or estimates possible. Costs will determine what is possible.

Take the case of open-pit coal mining. Lively controversies surround this procedure. Should we scar the earth and get the coal or save the hills and endure the cold? If the rights to open-pit mining are owned by a particular profit-maximizing agent, whether farmer or miner, that owner will begin to allocate the scarce rights to those who value them the most. It is pointless to consider the exact method he will use in the process, but we can be assured that the imagination of the owner will create a workable system. If he is in competition with other, similar owners, he is forced to create a system. It is not the method but the results that interest us.

**Conserving for the Future**

If an owner begins to obtain bids for open mining rights, those who believe that future generations will value them highly can validate their belief. They can put their money where their belief is and bid for the rights. If successful, the rights can be banked for future use and the hills saved. In fact, if the owner of the rights believes that future generations will place a sufficiently high value on the aesthetic beauty of the terrain, he will bank the rights himself. In making this decision, the discounted

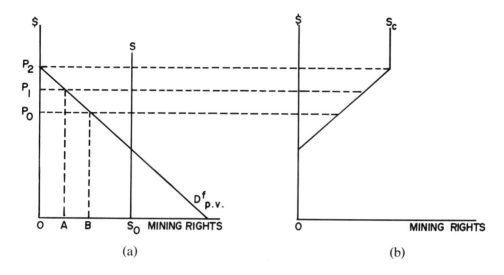

**Figure 6–1.** Analysis of Allocation of Mining Rights between Current and Future Uses

present value of the asset's future worth has to be weighted against the present value of some alternative use. It is a difficult estimation, but it is done every day at home as well as in the halls of finance. The manager knows the current market price of rights to land quality, so he can use that as his benchmark. It is possible that he may save an entire coal ridge. His analysis of the situation will determine that.

Figure 6–1 illustrates the situation just described. The relationships shown in (a) are those internal to the firm. They are as follows:

$S_o$ = the fixed supply of coal mining rights

$D^f_{p.v.}$ = the discounted present value of future demand for the rights, where

$$D^f_{p.v.} = f(P)e^{-rt},$$

which is a linear demand curve discounted at some interest rate, $r$, and for some time period, $t$.

Now, given a competitive market for rights, the owner will compare his reservation price, the discounted present value of future demand, with the price he can obtain in the current market for coal rights. For example, if the current market price is $P_o$, in figure 6–1, the owner will hold $B$ rights for the future and sell $S_o - B$ rights in the current market. If the current price is $P_1$, he will hold $A$ rights and sell $S_o - A$, and so on. By considering various current market prices, a supply curve of current rights, $S_c$ in (b) can be derived. At some price below $P_o$, an intersection between $S$ and $D^f_{p.v.}$ indicates that all rights will be held for the future.

The simple supply-demand analysis here shows that changes in the current price of rights could either override or enhance the tendency to save the asset for future use. Thus price becomes the determining factor. Current prices would reflect the needs of the present generation. The reservation price of the owner would reflect the needs of the future. For the sake of simplicity, we have placed the future in the hands of the single owner. Such will never be the case. Any number of bidders could consider the future; in fact, all bidders would. Each person would make the same estimates as the owner in our analysis. Even after the rights were distributed, the process would continue. Some would hold their rights for a time and then choose to sell them to others. Some would use their mining rights immediately. Thus, in buying and selling, both the present and the future would be considered.

Some people think the use of price is a heartless way to account for the future. But remember, the black hats and the white hats have an equal footing in the market. In fact, the hat that looks black to one person may look white to another. Society determines the value of the rights and the color of the hats. No one person or small group of people makes the decision. We should point out that the analytics described in figure 6–1 are actually a part of the analysis we showed in figure 3–1. There are several forces involved when one determines the marginal benefit or marginal cost associated with environmental use. One of these is the future.

### The Discount Rate

While both present and future values play important roles in the process outlined above, there is another important element—the discount rate.[2] Again, we must emphasize efficient markets. Individuals, real people, are making decisions. They are using scarce resources in the process. Information and time are conserved, along with every other scarce resource. The decision-maker is forced by competition to do the best he can. The discount rate will be chosen with care. What are the considerations?

First, the choice is affected by the alternatives faced by each individual decision-maker. These vary, but when adjusted for certain variables, they tend to converge as investments are made. For example, the owner of the scarce rights will probably choose a rate reflecting what he can earn from invested capital under similar conditions of risk. If his estimate of the future were perfect, without risk, if all calculations had been adjusted for price level changes, and if there were no taxes on the return on his investment in coal lands, it is possible that the interest rate used could be as low as two or three percent, a pure interest rate. Any risk premium, inflation effect, or tax distortion would have to be added to this figure.

To illustrate, if there were a two percent pure interest rate, a three percent risk premium, an expected rate of inflation of four percent, and a marginal tax rate of fifty percent, then the discount rate would be eighteen percent. After paying taxes on earnings, the manager would have nine percent. He would assign four percent to inflation and three percent to risk effects, leaving approximately two percent in pure interest.

Individuals not involved in a capital-conserving production process use similar logic in considering their bids for rights, whether to hold or to use. However, individual time preference is important here. Some individuals may consider the environment as a "special" investment. Thus, they might assign a high future value to the investment, indicating a special appreciation of future use versus present use. It is doubtful, however, that the environmental resource would have any more special following than other resources such as medical facilities, university libraries, cancer research foundations, or facilities to produce food and clothing. Investors tend to listen to the market, and they supply funds where the rate of return is high and the risk, low. Remember that society speaks of its concern for the future through the interest rate.

In addition to the chosen interest rate, time has a significant effect in any calculation of present value. This is particularly relevant when thinking of future generations. In that case, the time factor is large. To demonstrate this, imagine that a right to coal property will have a value of a million dollars in seventy-five years. The million dollars is the bidder's best estimate of future values. This must be converted to present value.[3]

Assume that the owner makes his calculation using three different interest rates, just for comparison purposes. The rates chosen are two percent, five percent, and twelve percent. The equation for calculating present value is:

$$P.V. = (F.V.)e^{-rt}$$

In our case, $t$ has a value of 75. The interest rate, $r$, will be filled by each of the three rates chosen.

After making the calculations the results are:

$$P.V._{.2} = \$223,000$$

$$P.V._{.5} = \phantom{0}23,500$$

$$P.V._{.12} = \phantom{00}123$$

Here we see vividly the effects of compound interest, a force often overlooked in discussions of future generations. The power of compound interest is coupled with a long planning horizon, and the effects are dramatic. We see the maximum amount that could be offered for a $1 million property right to be delivered in seventy-five years. If an alternative investment yields twelve percent, the value of that right in the present is $123.

In practical policy terms, if the owner of the coal rights had a choice between selling his coal for $123 today or holding it for seventy-five years and then realizing $1 million, he should be indifferent. If he could get more than $123 today, he should sell, for he would be better off and so would society. If the coal were used to produce more goods and services today than could be used to produce more goods and services in the future, at a twelve percent growth rate there would be more than $1 million in additional goods seventy-five years later. It may appear to some that twelve percent is an artificially high interest rate, and this is what stacks the cards against conservation. However, the average return on equity capital invested in manufacturing corporations over the ten years from 1964 to 1973 was 11.6% (*Statistical Abstract* 1974). Further, this return is calculated after corporate income taxes of about fifty percent. In other words, these firms earned between twenty and twenty-five percent on the equity funds invested in plant, equipment, and inventories. Investing today can yield large returns in the future, but this investing means using goods and services, and natural resources, to produce more goods for the future.

The significance of this approach to providing future benefits from an environmental resource can be appreciated if the concept of efficient markets is understood. The point is that society places many competing demands on its resource base. The interaction of markets in meeting these demands brings forth various investment opportunities. The rate of return available to investors is the signal from society that communicates intensity of demand. If investments with a low rate of return (low discount rate) are chosen in the face of higher returns elsewhere, society's signal has been disregarded.

Few people, if any, are capable of assimiliating enough information to discriminate subjectively among the myriad of social projects available to investors. For this reason, a discount rate based on alternative investments is accepted generally as an objective approach to the problem.

No matter what approach is taken in conserving environmental resources for the future, there will be a rate of return. There is no way to avoid placing a price on the environment. When one considers that approximately $152.7 billion will be invested in environmental control assets in the United States by 1981, it becomes apparent that an astronomically high value has been assigned to the future environment.[4] Even if the discount rate were two percent, calculations based on a seventy-five-year planning horizon after the completion of the ten-year program would show that future value would have to be almost four and one-half times the original investment. That isn't peanuts.

As an alternative to the solution that a free market may give us in providing goods for the future, we may consider some cases where government has sought overtly to provide for the future. There is good reason to consider government provision. Political realities indicate that the market will face tough competition from those who will vote for a centralized allocative authority when the future is considered. We have already noted that legislators and government administrators are reluctant to sacrifice any power that they may have. This applies to planning for the future as well as to providing for the present. In fact, there is currently a marked effort to have the government participate more actively in planning for the future.

One interesting example of the government's providing for future needs concerns energy. The principal natural sources of energy used in the United States are coal, oil, and natural gas. Government efforts to plan for future use have been quite limited in the case of coal; more active in the case of oil, with the "Hot Oil Acts" of 1933 and 1935, which assisted states in their efforts to regulate output and to avoide future shortages; and most active in the case of natural gas, with controls over price and conditions of transmission. If we consider today as the future that was planned for in these earlier government actions, we seem to have inherited problems in direct proportion to government action.

A more methodical study has been done by Gramm and Maurice (1975), and it offers some interesting evidence. Their work focused on fifteen depletable natural resources. Taking 1900 as a base year, they posed the same question we developed for the mining rights manager. Their question was specific and empirically testable. If the 1973 prices of the resources, deflated for inflation, had been known by government decision makers in 1900, would it have been in the best interest of society to stockpile any of the fifteen resources for future generations?

They made their calculation with a discount rate of only three percent. The discounted present values were compared with the market price of the resource in 1900, or as near to that date as possible. In every case, a decision to hold the resource would have been a loss for society. Natural gas happened to show the smallest loss for society, a loss of seventy-two percent of the investment!

For example, the cost of natural gas at the point of production was 8.3 cents per thousand cubic feet (mcf) in 1919 and 21.3 cents per mcf in 1973. It thus appears to many that it was wise to save the gas for future use. But

88

if the gas had been sold at its 1919 market price and the proceeds invested at three percent annual interest, by 1973 the value of this investment would have risen from 8.3 cents to 41.9 cents. If the natural gas had been saved, its value in 1973 would be just fifty-one percent of what we could have today if we used the asset to produce more goods and services. Stated differently, we could today have goods worth almost twice the value of the gas we hoarded—and with only a three percent return!

Of course, past government action has resulted in petroleum, natural gas, timber, and various metals, being withheld from the market. Although this action cannot be supported on the basis of economic analysis, it has occurred. So what difference does it make? As indicated earlier, diversion of resources from the market when scarcity is already present means that some socially useful activity will not take place. In the cases cited, we gave up using oil, copper, nickle, and other resources to produce schools, mills, machinery, roads, drugs, and the like so that these natural resources would be available in the future. Starvation of the present will bring a distorted future. And who can say what might have been sacrificed? By holding goods off the market, a clean environment along with many other goods became more costly. The evidence is clear in the study reported. One might summarize by using the old saw, "Where were you when I needed you most?" Efficient markets, though not perfect, tend to react quickly to society's demand.

### Experience under Government Ownership

Sometimes it is helpful to search for experiences similar to that which might become policy when "new" economic problems are considered. The word "new" is in quotation marks for quite obvious reasons. Even an amateurish review of history quickly confirms the notion that there are few new problems. Public ownership of industries closely associated with natural resources has been the policy in Britain since the late forties. As reported by William G. Shepherd (1965: 6), "The Labour Government established the National Coal Board in January 1947, the British Electricity Authority in April 1948, and the Gas Council in May 1949, thereby extending nationwide public ownership to virtually all of the domestic energy sector of England, Scotland, and Wales."

It must be recognized that these industries were regulated prior to their nationalization. That is, the market for energy-related products, even prior to public ownership, had not been free in the efficient markets sense of the word. Nonetheless, the changes wrought under governmental ownership were significant. In discussing the problems and peculiarities of these industries, Shepherd does not take a position favoring a return to private ownership. He expresses concern for monopoly power that could result.[5] He does, however, suggest commercial behavior for the governmental owners, implying that market forces would be beneficial in rationing capital, bringing competition, and generally orienting the managers of these industries toward an efficient market solution.

The British experience with natural resources is a record of public ownership with little regard for the signals of the market. The approach used in other times and places indicates that the market can improve public enterprise as it has private. An interesting example of this involved the introduction of parking meters in the United States. Prior to their introduction in the mid-1930s, parking was free but often unobtainable. When Oklahoma City and Dallas first began using parking meters, certain changes occurred. People looking for places to park found them. Owners of stores and their workers found other places to park all day without taking up valuable street parking. Less damage was done to cars going in and out of parking places, because parking places were enlarged and clearly marked. Business firms in the metered areas found their business increased because people now felt certain of being able to park when they came to the downtown area. Finally, the city gained revenue. All the advantages predicted from pricing appeared: more efficient use of the space; use by those who valued it most; and improved social well-being (Macaulay and Yandle 1975). Thus, public enterprise can use market methods to manage scarce resources.

**Environmental Safety and Economic Efficiency**

Robert Burns has warned us about the best laid plans. There are accidents with which we must contend, random events that create unusual costs. In some cases man is the victim, and the environment, the culprit. Then there are times when the situation is reversed. Man may go merrily along spewing out smoke daily in large and small quantities, but he may also produce a Santa Barbara oil spill. Nature does some interesting things alone. Life may be pleasant and rather calm, but then volcanoes erupt, hurricanes come, tornadoes strike, and tidal waves sweep the land, as do earthquakes, droughts, and locusts. There are numerous forms of pollution.

Still, whether the accidents are man's or nature's, both can be handled in a market context. The mechanism is insurance, and property rights are critical in its functioning. For example, the first recorded insurance policy dealt with accidents at sea.[6] But the asset was the cargo, not the sea. The sea was the hazard. Where shipping activity developed, insurance became a growing commercial enterprise.

The mechanism can be described simply; however, in actual practice it is rather sophisticated. Simply stated, where a large number of people confront a similar hazard, a long-run expected loss is predictable. The law of large numbers does it. Historical records enable a manager of risk to estimate the expected damage, to calculate a pro rata cost, and to invite people to pool their risk. Since any one individual cannot estimate his future loss with great precision, it pays him to pool his activity with that of others. Some will pay more than they gain; others will gain more than they pay. But taken together, the total payments cover the cost of living in a hazardous environment. Insurance spreads the risk and provides for an uncertain future.

Much of man's environment is insured in just this fashion. The risks of flood, fire, earthquake, occupational damage, loss of life can each be insured. The insurance mechanism is the market solution to the problem.

By thinking about the natural environment as an asset, we see that it, too, can be insured. Three economists reviewed experiences with oil spills by tankers and made just such a proposal. Their findings are interesting.[7] It seems that some 2,749 accidents involving oil spills occurred along the coast of Great Britain between the years 1959 and 1968. It is also estimated that some 200,000 tons of oil spill into the world's seas each year from tanker accidents. Lest the reader take alarm and head for the nearest mountain, we hasten to point out that natural seepage in seas exceeds this amount. Man is a minor polluter.[8]

There is little doubt, however, that some economic damage is done to the natural environment. Whether the cost to society would be greater or less after taking steps to avoid the damage is a separate and far more relevant question. Nonetheless, sandy beaches are temporarily blackened, fisheries are reduced in value, holiday resorts suffer for a day or a season, and real property may lose value, all as a result of environmental change due to man's accidents at sea.

While we are discussing oil spills such as the one that occurred at Santa Barbara, this is perhaps a good time to test the reader's grasp of the case that has been made for charging people who use the environment. Here is the question: "Who should pay for what happened in the Santa Barbara Channel?" The oil companies drilling there could have achieved varying levels of pollution abatement by paying increasing costs, while residents in the area would have valued increasing levels of purity at less and less per unit of purity. Think about our tried and true friend, figure 3–1. If either the government or some private individual owned the Santa Barbara Channel, he would charge both oil companies and residents along the channel for their particular use of the asset. Thus, although some oil would have flowed into the channel, residents along the shore would have had to pay. Although such a policy would neither win friends nor influence people on shore so long as they thought they could get complete cleanliness for nothing, it would be the best use of the channel. Oil firms might also frown on the solution. They would have to spend money to prevent spills and still have to pay the channel owner, even if they spilt less oil than previously. The hurdle is seeing that residents must pay though they still get oil on troubled waters. What they pay for is less oil on the water.

The new environmental economics, which suggests that everyone owns everything, or that everything owns everyone, makes this solution sound heartless. But recall the object of economics. In the face of scarcity, we would like to bring about asset utilization that leads to welfare improvements for man. Those who get cleaner channels pay for the benefits. Those who get additional oil, pay their costs also—all of them.

## The Insurance Question

As we have pointed out, a private market can operate once property rights are established. The same can be said about insurance. In the case of oil spills, one could imagine the owner of sea lanes and unloading docks charging for their total environmental use. That is, the owner of the facilities could establish a fee to include the expected cost of cleaning up and reimbursing adjacent environmental property owners for the unanticipated use of their property, or he could pay an insurance premium to cover the same eventualities. In either case, price would communicate scarcity.

In a setting of full property rights to the scarce environment, one pictures an interesting result (Burrows, Rowley, and Owen 1974). When oil terminals are located in hazardous places—places with a large risk of environmental damage—the fee for use would be high. In remote, less hazardous places, the fee would be less. Shippers would respond to price, another cost of carrying oil, and select the least-cost alternative. By competing with other port terminals, the individual operator would seek to make his terminal as environmentally efficient as could be financially rewarding. He would want his fee to be attractive. Perhaps he would lower the charge for nonspillers, tankers with exceptionally good records of care, and raise it for the spillers. Differential prices would spur ship owners to use the most economical equipment. By conserving their cash, they would also conserve the environmental asset at an optimal level of purity. Property rights and insurance would do the trick.

There is more to this approach than first meets the eye. In order for the sea-lane operators to set their fees, the ownership of environmental rights in every competing neighborhood must be established. That is, they must be scarce enough for someone to own them. Only the owner could be paid for damage, not every person seeing, hearing, or smelling the accident. Just as with an automobile accident, only the damaged party, the owner, is made whole. If automobiles were common access resources or publicly owned, the number of damaged parties could become almost infinite. Each person might feel that his transportation environment was reduced if an automobile were struck from the fleet. Following this thinking, each person might demand full compensation. It just won't work. We shall develop the point further.

Unfortunately, it is the latter view that certain proponents of environmental insurance have in mind. There are other problems that can also creep into the insurance scheme, problems associated with the politico-legal environment. Let us see what can happen to a market instrument when it becomes "public."

Automobile liability and medical malpractice insurance raise a disastrous specter. Part of the problem stems from the new environmental way of looking at things. This is seen in the open-ended nature of claims, the unsettled value associated with damages. Only the imagination of the claimants and their attorneys seems to limit the amount of the damages sought. Judgments by juries seem to have no upper limit. As a result, payments skyrocket, insurance premiums follow the same pattern, and prices charged by those who offer transportation and medical services do the same.

The public nature of the claim settlement seems to be the culprit. Defining the value of the damaged asset is not well established. As a consequence, no-fault insurance enters the picture. The question of liability is skirted, and the limits of payment are set in the contract. Unfortunately, the maximum limit tends to become the amount paid.

A second solution comes quickly to the front. Let the government insure. If insurance companies cannot keep up with ever-escalating claims or if doctors and taxi drivers cannot sell their services at prices high enough to cover their insurance premiums, it is necessary to have an insuror of last resort. Then the entire process becomes disguised. We are back where we started. Tax revenues are used to reduce deficits that result in the insurance programs. Premiums tend to become meaningless in terms of communicating scarcity, and the all important impact on human behavior is reduced. Individuals tend to resume their hazardous ways. The total environment becomes polluted with risk.

The final solution to this problem rests on the suggestion we have already made. Property rights must be defined where there is scarcity. These rights must be purchased by those who place a value on them. With an explicit market, prices will exist. In turn, prices will set values. And values determine the amount of damages that can be claimed. If we have markets for all environmental goods, the problem of risk is solved.

### The Future of the Market

We have now discussed various ways in which the market handles the future. First, we described the process in which investors consider the future when determining the price they will pay for an asset. Then we provided a brief discussion of government and its approach to the future. After that, we offered a summary of insurance and how future risks are accounted for. Now we turn to the market process itself. How bright is its future? We cannot answer the question, for two reasons. First, we must ask, "Relative to what?" The answer to that question may give us something to go on. But there is a deeper problem. The market is truly all-pervasive. We believe it would be impossible to determine just what its state of health may be. However, there are some problems that we shall mention here.

When one talks about the market, predictably someone will say, "But markets don't really exist. They are just a part of some economist's theoretical tools." This is a communication problem.

Economists stress certain words when they talk about markets, supply curves, and demand curves.[9] Some of the words cause the problem reflected in the criticism voiced about theoretical tools. The words "pure or perfect competition" have a very restrictive meaning in economics. There is a structural emphasis: many buyers and many sellers, none large enough to affect price, all engaged in voluntary exchange. There is also a qualitative emphasis: perfect mobility and knowledge, a frictionless flow of information, movement with ease into and out of the market. Any buyer or seller can move instantly from one transaction to another, always understanding the price conditions encountered in the market.

In addition, the items traded are described carefully under the pure competition rubric. Each good in a particular market is viewed by the buyer as a perfect substitute for any other good in that market. The world may be described as if it were composed of homogeneous, look-alike products, all traded by utility-maximizing, all-knowing, high-speed buyers and sellers.

As the theory goes, if any of these assumptions is compromised, another word takes over the market. "Monopoly" is used to describe the absence of pure competition. Its use implies a loss of randomness in the marketplace. An observer could then make accurate predictions of which seller may sell to particular buyers. An element of control enters the picture. The entry of control implies a dramatic change in the social benefits of market allocation. Whereas pure competition sets up conditions leading toward least-cost production, maximum satisfaction for consumers, and the rapid translation of consumer demand into production, monopoly shifts the benefits to one side of the market. Whoever holds the power will seek predictably to gain as much as possible for himself. Price tends to rise, output tends to fall, and total satisfaction tends to be reduced.

The words "pure competition" and "monopoly" are extended beyond consumption to production. In the world of pure competition, each producer uses the same production techniques, the same inputs in the same proportions and at the same cost. Conditions other than these bring forth the word "monopoly," with its distorting effects.

This off-on switch of competition and monopoly gives us a set of conditions that provide analytical power. But the use of the words, the switch, is often misunderstood. Because none of the conditions described under pure competition seems to exist in a real sense, critics wonder about their usefulness. If the conditions do not exist, does that mean that the whole world is made up of monopolies? If that is the case, why would anyone in his right mind argue for allocation by the market?

Well-developed theories of economics, such as pure competition and monopoly, bring a common problem. The restrictive assumptions are compared with the real world, testing in a sense the possibility of obtaining the results. Of course, the results of pure competition are attractive. With such massive social benefits in the offing, it is tempting to search for those conditions that satisfy the theory, or at least to change the world, if possible, to meet the requirements of the assumptions.

Some policy makers, fearing the possibility of a world filled with monopoly power, argue for many small buyers and sellers, competing in a single market for products wrapped in plain packaging so as to prevent product differentiation, and for a ban on all advertising except that which is merely informational. By meeting the restrictive assumptions, they hope that the world will reach new and higher levels of happiness. To fail brings on the horrors of monopoly power.

Other trappings of economic theory have equally frustrating policy implications. The now well-developed theories of externalities and public goods have been called to the attention of practitioners in every field. Attorneys, businessmen, planners, political scientists, engineers, and sociologists are all aware of market failures. Many have grasped only that much of the theory. They have to realize that markets, like the rest of the world, are not perfect. There are side effects, spillover benefits and costs, generated by any productive activity. The economic relevance of these costs and benefits is not pursued by people who see only markets that fail to be perfect. They see externalities that will not go away completely, for so long as any third party is aware of activities or transactions taking place around him, he will probably approve or disapprove of the results. Anything other than complete indifference could be called an externality.

The awareness of externalities produces new attacks on the market mechanism, attacks added to those by people who feel that pure competition, with all its welfare benefits, never existed in the first place. The externality critics add further injury to those who value economic freedom by asserting that markets can never allocate efficiently because of spillovers. It must be realized that there is no such thing as a perfect market. Nor is there a perfect environment. But there can be efficient economic markets and productive environments.

Sometimes it is beneficial to stop using certain words in an effort to circumvent theoretical tigers. Pure competition, with all its elegant *ex ante* assumptions, may be a tiger that stands in the way of clear thinking about markets. It is a tiger to the extent that misunderstandings occur. There are no fallacies in the theory itself.

A substitute term is now in use to describe an economically efficient system. The term is "efficient markets."

1. Goods are produced by as many producers as necessary to obtain least average cost of production. Because of the threat of entry from outside an industry, and the existence of rivalry within the industry, firms tend to price their output where marginal cost is equal to price.

2. Firms in the same industry use the most efficient production techniques available to them, all costs considered. Thus, it is not necessary that every firm use the identical technology, only the least cost technology.

3. Input factors of production, land, labor, and capital are hired from the cheapest source, all costs considered. Some firms pay higher wages, but spend less time in the search for inputs. Others hire less qualified workers and spend more time in training them. Similar firms competing in the same area do not necessarily have identically structured labor forces or uses of capital and land. They do however, tend to have similar average and marginal costs of production.

4. Consumers of products shop for goods that offer them the least-cost package of benefits, all costs considered. Just as with the firm, some search longer than others, some use more low-quality goods, but all tend to behave efficiently, given the costs associated with changing their behavior.

The law of efficient markets follows from the above descriptive assumptions. Nothing in them implies perfection or purity. Everything focuses on costs. Where there are open markets, economic freedom, and private property rights with implied enforcement of contracts, costs will be forced to a minimum at every level of decision making.[10]

By following this line of logic, we can return full circle and perhaps appreciate better the pure competition paradigm. Now it must be realized that pure competition is an *ex post* condition, not *ex ante*. That is, given efficient markets, as outlined here, the resulting equilibrium can be described *as if* many buyers and sellers were exchanging a homogeneous product in a well-informed manner. Relevant externalities are negotiated away where possible. Those external effects remaining are tolerated *so long as* the market mechanism and its original system of property rights remain. As discussed in chapter 5, forces can and will be set in motion to develop new rights to scarce property if externalities are present. Eventually, the costs of making such an institutional change could be covered by the benefits.

While controversies still rage in the externalities camp, the notion expressed here has been recognized and articulated many times.[11] Perhaps J. G. Head (1974) first discussed completely market efficiency in a real world context. His discussion included some of the costs associated with dealing with externalities that have been touched on here. However, he went on to discuss another equally important cost—"The Concept of Political Costs."[12] In Head's summary of the problem, he raises this very question:

Where externalities are observed and marginal social net products are not equalized (imperfect markets), public intervention may be possible to improve resource allocation—or it may not. Such intervention inevitably involves a variety of "political costs," and is therefore justified on efficiency grounds only when these costs are outweighted by the allocative benefits (p. 211).

The efficient conditions, the maximization of society's well-being associated with efficient markets, underlie all of this discussion. We shall not review these conditions; they are well-known and a part of the memory banks of nearly every student who has taken a second course in economics. But we shall emphasize the underlying requirement of voluntary exchange, and its purpose. Sometimes this becomes lost in the more technical discussions of economics.

Voluntary exchange takes place because individuals believe that they will be better off after the exchange than they were before. It is the pursuit of happiness, not goods, that brings about the desire to exchange products or services. Efficiency and efficient markets refer to the result obtained by the unrestrained action of individuals motivated by legitimate self-interest.[13]

The use of the word "efficiency" has itself created some confusion. Some think of the word in physical terms: the greatest physical output for the smallest physical input. This presents the danger of confusing countable things with happiness. The conditions necessary for voluntary exchange will not assure necessarily the greatest physical output of any product. But there will be an assurance of the greatest happiness, something that cannot be ordered up with legislation.

## Efficient Markets: Lessons from History

To a great extent, acceptance of the market is a matter of faith. We have indicated that one cannot estimate the value of the environment, but that it is done. We have said that it would be futile to attempt an explanation of how a coal-rights manager might manage his firm to achieve an efficient use of his assets, but that it would happen. Our effort to describe the concept of efficient markets has also turned on the expected benefits of voluntary exchange in a competitive drive to use property rights. At every turn, it seems, we have implied that, left to itself, the market handles many problems in an efficient, socially beneficial manner.

Part of this belief in markets is based on substantial theoretical study and countless reports of empirical research by economists. Another element comes from a knowledge of history. There is a basis for hope in the pages of history.

To illustrate, let us review two historical lessons dealing with natural resources. They relate to energy. In fact, modern journalists would probably describe these two situations as energy crises. But there was no Federal Energy Office to beat the drum when these crises occurred.

It is a well-established fact that America experienced her first energy crisis when whale oil, a principal source of lighting, became scarce (Wriston 1974). There had been cartel attempts on that source of oil also, and John Hancock, the revolutionary hero, had his ins and outs with the early monopoly effort (Gras and Larson 1939). Eventually, however, a growing demand and a shrinking supply pushed the price of sperm oil to the heights. In fact, the price of the product doubled during the course of the Civil War.

The Law of Efficient Markets suggests that a rising price triggers a social response. New producers enter the indicated lucrative trade and other entrepreneurs find it profitable to search for substitutes. The rising price communicates all of this. And the market was working in 1849, for the response came. In fact, a totally unexpected, yet highly predictable, result obtained. The wisest man never would have expected what was to happen. Yet it could be predicted, based on thousands of historical examples. From the introduction of coal to the invention of the transistor, "Necessity is the mother of invention."[14] Yet, "wise" men facing the same problem today would probably impose price controls, order a fifty-five grams-per-hour limit on use, and set up storage vessels for whale oil in order that future generations might not curse the darkness.

The solution to the problem began in 1849, when Samuel M. Kier, a salt merchant, began to market a medicinal oil as a sideline to his business.[15] The oil, a nuisance then, had been found in salt-water wells. To merchandise the product, Kier had an artist design a fancy label for the bottled product. On the label was a sketch of a salt-water derrick. All the while, the price of whale oil was rising. The market was sending its signal.

A young professor at Dartmouth College saw one of the bottles and had an idea. George Bissell must have thought, "Why not drill for oil?" With J.G. Eveletz of New York City, he organized the Pennsylvania Rock Oil Company. And with their total capital of $5,000, they purchased what they thought were the property rights to underground oil in Pennsylvania.

Lacking established expertise in drilling, but believing that they would become rich, the proprietors of the new firm hired E.L. Drake, later called "Colonel," and had this former railroad conductor take a salt drill with him from Hartford, Connecticut, to Titusville, Pennsylvania.

After much difficulty, they sank a well thirty-five feet deep. They struck oil in 1859. Despite the energy crisis, there was opposition against this unholy work. A local preacher condemned the project as immoral, because, he said, the oil was needed down there to feed the fires of hell.[16] To withdraw it was to protect the wicked from the punishment they so justly deserved!

There was already a growing market for petroleum products. Some 2,000 barrels of oil were produced annually in the 1850s from salt wells and rock seepages. Drake's well produced that much the first year, and his example was followed vigorously. By 1869, there were 4,800,000 barrels a year coming forth. Efficient markets were working. A new industry had been born and the pressure was taken off the whales. But with the profits from petroleum, there was an incentive for other competition. That came in the form of the electric lamp. The first practical generator for producing the new energy for lighting was installed in 1875. By 1896, the price of whale oil had dropped to $.40 a gallon from $2.55 a gallon in 1859 (Wriston 1974: p. 644). The market can do some impressive things when left alone.

**Another Energy Crisis**

There is another story dealing with energy that is at least as impressive as the one about oil, if not more so. It deals with an event that occurred in 1776, an event that changed the world. In fact, the change was referred to as an industrial revolution. In 1776, the first Boulton-Watt steam engine was installed and placed in productive use (Gras and Larson 1939: 190–209).

There is a long story behind this achievement, but a few events suffice to describe the operation of the market at that time. England was running out of wood, the primary source of energy in homes and factories. Some people feared that they might freeze, and others were threatened with plant shutdowns and unemployment. Of course, there was coal, but it was very expensive to mine, and there was no reliable, inexpensive way to move it from the mines in Wales to the people and plants. A serious problem in the coal mines resulted from the collection of water in the tunnels. New pumps had been invented, but there was only horsepower to make them work. Things looked rather bleak.

James Watt was a young technician employed by the University of Glasgow. While repairing a Newcomen steam engine at the University, he discovered a method to improve it. Somehow, Watt's engineering expertise came to the attention of Matthew Boulton, a successful machinery manufacturer near Birmingham. Boulton saw some financially attractive possibilities in the work. Once again, the market was getting the message through.

The contents of a letter received by Watt (Gras and Larson 1939: 191) gives an indication of the situation:

A friend of Boulton and me in Cornwall sent us word four days ago that four or five coppermines are just going to be abandoned because of the high price of coals, and beg me to apply to them instantly. Yesterday application was made to me, by a mining company in Derbyshire, to know when you will be in England about fire engines, because they must quit their mine if you cannot relieve them.

Boulton pawned his entire fortune and placed it on a bet that Watt's idea could become a commercial success. The two formed a partnership that even today would inspire envy in an Ayn Rand entrepreneur. They produced, sold, and installed engines, for profit. In spite of their efforts to hold on to their patents, the law of efficient markets went into operation. As the word spread about their financial success, more and more inventors entered the race.

Another inventor, Richard Trevithik, finally decided to mount the high-pressure steam engine on a tram car: this was the first locomotive to carry coal from the mines to London. The Boulton-Watt steam engine ran the pumps that made the mines functional. The Trevithik locomotive carried the coal. All without national planning, Congressional hearings, or bureaucratic regulations and guidance.

Just these two vignettes from history may serve to make our point. Free men motivated by legitimate self-interest will seek to improve their own lot and in so doing will provide good and goods for mankind. Today, people may suggest that things were simpler in 1859 or 1776, that the complex situation we face now cannot be compared with those of Kier, Bissell, Drake, Watt, and Boulton. This can be said now, but could it have been said then? Somehow, many generations prospered beyond the imaginations of those who faced the early energy crises. Somehow, markets were efficient to the extent that resources were conserved, new technologies developed, impossible engineering feats accomplished, unheard-of substitute resources tapped, and the welfare of mankind improved. What more could one look for in a brief history lesson?

**Notes**

1. In referring to the tendency to look to the future in providing water, Boulding says, "There are some places, perhaps in the arid areas, where too little investment is going on. There are almost certainly other places, such as Los Angeles, where too much has gone on and is still going on. Los Angeles is going to run out of air long before it is going to run out of water. This is almost a classic example of economic presbyopia—farsightedness in the optical sense of the term. In Los Angeles water is not a commodity but a religion" (1964: 88).

2. There are few subjects more discussed than the appropriate discount rate for use in public investment analysis. Extensive work has been done in the area by Arnold C. Harberger (1972), Charles W. Howe (1971), and the United States Congress Joint Economic Committee (1969).

3. Floyd E. Gillis (1969) provides a very readable discussion of the concept of discounted present value, especially in Chapters 2 and 4.

4. United States Council on Environmental Quality, 1973. Estimates of the cost of pollution abatement show a wide range. Thus, while the CEQ estimated that $153 billion will be invested in environmental control assets by 1981, the National Water Commission estimated that capital investments of $371 billion would be needed between 1973 and 1983 just to meet the "best known technology" requirements of the Federal Water Pollution Control Act Amendments of 1972. This latter figure applies only to water, does not include $97 billion of estimated operating and maintenance costs, and is expressed in 1972 dollars (p. 513). Kneese and Schulze (1975: 54) deal with the zero discharge requirement of the Act by assuming it is so ridiculous that it will never be enforced. But it is the law.

5. After studying the electric utility industry, George Stigler and Claire Friedland (1962) concluded that even where a natural monopoly existed, its prices were not significantly different from those charged by a regulated monopoly. If monopoly power does exist, government regulation does not seem to abate it.

6. The first policy was dated February 13, 1343, and covered a cargo ship sailing from Genoa, Italy. Earlier efforts to insure amounted to loans between merchants and ship captains. These go back to 2000 B.C. (Nelli 1972).

7. See Burrows, Rowley, and Owen (1974) for an interesting analysis

of the Torrey Canyon grounding, a tanker accident that caused the loss of 119,000 tons of crude oil.

8. Nature has not only a way of diluting much of its own pollution and man's but also a way of gradually absorbing and adjusting to what we call pollution. Thus, *Clean Water Report* tells of a research team headed by a biological sciences professor at the University of Southern California who "has found that the 1969 oil spill in Santa Barbara Channel did almost no permanent damage to animal and plant life or to beaches, and that the area has almost completely recovered." The team suggested that one reason for the limited damage to plants and animals was that they had adjusted to constant exposure to small amounts of oil in the water due to natural seepage. "It was found that areas which regularly receive oil, have abundant inter-tidal flora and fauna" (February 1971: 15).

9. In his *Theory of Moral Sentiments*, (1966 ed.) Adam Smith discussed words as abstractions and theorized about the development of various parts of speech.

10. Demsetz (1973: 26–27) provides an excellent discussion of the problem with the words "competition" and "monopoly." He says, "But these assumptions are properly treated as thought-facilitating devices and not as description of the structure of a real monopoly but as a proxy for the statement that 'for the purpose of the problem at hand, competitive behavior can be ignored.' . . . Similarly, the many-firm assumption of the competition model is to be thought of as a proxy for the statement that 'for the purposes at hand, monopoly power can be ignored,' rather than as a description of the structure of the competitive industry."

11. For a careful delineation of externalities relevant to markets, see Buchanan and Stubblebine (1962). Their analysis gives mathematical rigor to the classic work by Coase (1960). An emphasis treating explicitly the cost of markets is given by Demsetz (1964).

12. See Head (1974: 184–213). This section in the volume is a version of an article first published in *Rivista di diritto finanziario e scienze della finanze* (pp. 384–414). Also see McKean (1972).

13. The words "legitimate self-interest" are taken from the persuasive writing of Frederic Bastiat (1964: especially xxi–xxxvii).

14. It should be remembered that Veblen's observation that "Invention is the mother of necessity" is a complementary, not a substitute, rule.

15. For details on much that follows, see Boorstin (1973).

16. See Peach (p. 6).

# 7

## Errors in the Economic Theory of Environmental Use

We have seen that the environment can be treated like any other economic asset; if we make that simple move, the establishment of property rights in a free market can lead to an efficient solution of the problem of use. Yet the economic treatment of the environment developed here, to follow those precepts, has been a theory apart. Apparently, there are four factors that explain why older theories of economic use have been forsaken and new pollution theories adopted. All four factors result from errors in economic thinking. First, there has been concern about pollution of the environment rather than efficiency in its use, and the accepted definition of pollution has been so limited as to produce inefficiency in use. Next, many people have worried about compensating parties who were damaged by pollution when they should have been concerned about compensating the owners of assets that were reduced in value. Indeed, compensating damaged parties may be not only physically but also financially impossible because of errors in defining damage. Next, the environment has been viewed as a public good that can be enjoyed by many people simultaneously, while the basic problem faced is scarcity and the allocation of the environment as a private good. Finally, the spillover effects of environmental use have been observed with such attention that a concept of technological externalities has been established, with an extensive literature to support the subject. The older concept of pecuniary externalities was adequate to embrace and analyze the problem.

The four-part attack was more than the discipline could withstand. After the assault came a vast literature prescribing inefficient solutions to pollution. We shall look at each of the four sources of error to see how both simple and complex ideas can be misapplied to problems, so that the solutions create even worse problems.

### What is Pollution?

In chapter 1, we talked briefly about the concept of pollution and pointed out that even the concept is beset with problems, but that for our purposes we need a definition that is useful in dealing with economic questions. However, people tend to define pollution in physical terms rather than in economic terms. As noted in chapter 6, a similar problem arises with the definition of efficiency. Many will recall that early lesson in economics textbooks dealing with the difference between mechanical efficiency and economic efficiency. A modern reader may point out that atomic fusion is one of the most efficient ways to produce electricity from the standpoint of energy conversion; but cost factors at the present time limit this process to the laboratory for brief periods measured in microseconds and at costs of millions of dollars per kilowatt hour. We must add utility to the cost element. Even if plastic houses prove to be less costly than their brick counterparts, the pursuit of happiness may cause buyers to choose brick houses. Happiness and costs prevail in economic analysis.

Economic considerations are important in defining efficiency when the question involves producing or using. They are also important in defining pollution when we want to know what to do about it. However, physical measures of pollution have prevailed. Despite this fact, conflicts persist. If some people support the draining of swamps as good in physical terms, others scream "Pollution!" The economic problem must be settled no matter how one measures efficiency. And economists must use a definition that best suits their purpose. Ecologists and other scholars may employ concepts that do the most for their analyses. Communications across disciplines will be difficult, but mistakes internal to each discipline may be minimized. That would be progress.

It has been suggested that whenever pollution has economic consequences, these arise because some feature or service of the environment is scarce. That is, more than one user seeks to enjoy some of the benefits of the environment that are to some extent mutually exclusive. Thus, we have proposed as a definition of environmental pollution, any use of the asset reducing the benefit that any other person would otherwise enjoy from the environment. Though such a condition clearly has economic significance, it may represent a considerable departure from what non-economists consider as pollution.

Take the extreme case of a person who wants water in a stream to remain in its natural state, unsullied by any action of man, although the presence of a few leaves from trees and deposits from wildlife is tolerable. Suppose this person plans not to go near the water but only to look at it from afar or only to conjure up a picture in his mind's eye and enjoy its beauty. Is this pollution? Clearly it is not in the usual or physical sense of the word. But if he gets his way by the force of law, in an economic sense it is pollution just as much as if a resident upstream got his way by gravity as he put sewage and mercury in the stream to get them off his property.

For an economist who must analyze the effects of environmental pollution, there is a symmetry in the two cases that permits one theory to deal with both cases. To the extent he gets his way, each user may limit the use of the stream, or the environment, by the other. If the economist uses the physical concept of pollution, he may well devise a separate theory to deal with pollution when none is needed. This, in fact, is what most economists have done.

Sometimes it is instructive to review the evolution of ideas. At different times, prominent economists have written on the subject of pollution. For example, Baumol points out in his case of smoke-producing firms and laundries, "The laundry whose output is damaged by smoky air does not, by an increase of its own output, make the air cleaner or dirtier for others" (1972: 311–312). This is true but irrelevant, if not misleading, for economic considerations, for Baumol proposes that the smoke producer reduce his output so the laundry can increase its output.

A more reasonable proposal could stress a different term or idea, such as, "the user of the environment should pay," or, more completely, "the person who enjoys the use of, or benefits from, the environment, at the expense of some other user, should pay." The word "user" in the first definition may be considered the same as "polluter", and only uses that pollute, in the popular sense of the word, may be seen as true uses. The second definition is more inclusive, and still has a ring of logic and justice about it.

Each user whose use restricts the enjoyment that some other user would otherwise experience is a polluter. Thus, conservationists are polluters. And "the polluter should pay."

## Compensating the Damaged Party

Economists who have argued for charges have often argued also for compensation to flow from the polluter to the damaged party. For example, Musgrave has observed (1959: 45) "The factory that causes a smoke nuisance may be required to raise the height of its chimney or to pay damages in the form of a tax used in turn to reimburse those who have suffered in the process." In our arguments above, we have pointed out that all competitive users pollute, and practically all users are damaged. Only if one party has his way without limit does he escape damage. Traditional price theory calls for users of a scarce resource to pay for the use they enjoy. There is no reason to abandon this theory or to change its application when the environment is the asset involved.

Under this theory, charges are levied on the user of an asset to ration its use. Payments are made to the owner to assure its optimal allocation, and, where supply is not perfectly inelastic, to create an appropriate supply. If an individual or firm takes and uses units of some asset, to the exclusion of other users, the owner of the asset cannot use these units himself or sell them to others. He is thereby damaged and should be compensated for this damage. This is merely a simple application of the concept of opportunity cost.

In the case of the environment, two conditions exist that may lead to a misapplication of these general policies. In the first place, the supply of environmental quality in the natural state is considered fixed and not subject to an increase in quality at higher prices. Although at sufficiently high prices, improvements in air and water quality may be provided by man, these are limited sources compared to what nature provides. As a consequence, there is often little need to compensate owners to elicit increased supplies. In the second place, there have been no widely recognized property rights to the environment. That is the problem. Thus, the party most obviously damaged was the party who sought a competing use but because of gravity or wind currents was denied his intended use.

Because of these forces and the failure to perceive that "pollution" is actually "use" and has a bilateral or multilateral nature, most writers have called for compensation for the injured or damaged parties, but not for the owner. Buchanan and Stubblebine (1962), who come closest to recognizing the bilateral nature of externalities, reverse the direction of compensation in the example they cite. After they have been careful to give the acting party, in the popular sense the polluting party, the legal right to engage in his activity, they state that "the externally affected party must compensate the acting party for modifying his behavior." Still, their description states that "the acting party is being compensated for 'suffering' internal economies and diseconomies, that is, divergences between 'private' marginal costs and benefits, *measured in the absence of compensation*" (p. 381). One can think of the compensated party either as an owner deprived of the use of his property or as a user of the environment deprived of his intended use by another user. The fact that the owner is also a user in this example clouds the issue and the solution.

The appeal of the idea of compensating sufferers has been so strong that it has at times been applied to cases that are fundamentally different from those just cited. Thus, in another case, Musgrave deals with the short-run problem of crowding or congestion in the use of a bridge. "If there is [crowding], an auction is in order to ration out limited space. That is to say, tolls may be charged to those who wish to use the bridge without crowding, with the proceeds paid to those who in return are willing to stay away" (1959: 138). In the absence of property rights, one can hardly argue that potential users are owners who are entitled, for that reason, to payment. Rather, compensation is justified here because of the damage, or the reduction in welfare, that potential users suffer from their inability to use the bridge and their willingness to stay away.

However, this treatment is not normally adopted in rationing the use

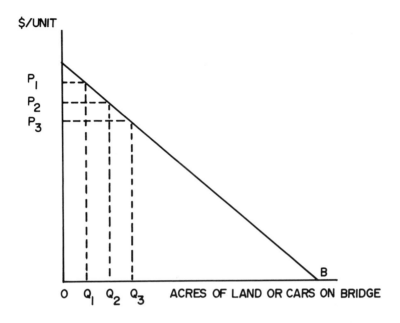

**Figure 7–1.** Demand for the Services of Land or a Bridge

of potentially congested bridges or highways, and with considerable justification. It will help us in our understanding if we return to some solid ground—land. Assume a two-acre site that is sought by 100 potential demanders who each want one acre and who, when ranged from the highest to lowest bidder, would bid prices per acre successively lower by one dollar from $100 at the top to one dollar at the bottom. We would then have a (smoothed) demand curve for land as shown in figure 7–1. If $Q_1$ represents the two acres bought by the two highest buyers, that is the amount that will be sold at price $P_1$ of $99. All other demanders will want land, but they will not get any, because it went to the highest bidders.

Now return to the bridge. Assume that the same demand curve shows the demand by people for the services of the bridge, and that congestion sets in at $Q_1$, traffic flow is at the optimum at $Q_2$ and that traffic ceases to flow at $Q_3$. It could well be possible that at either $Q_1$ or $Q_2$, the amount of revenue collected, even with first-degree price discrimination, would be too little to compensate all other potential users. That was certainly true in the land example used. If the amount required for compensation to nonusers were higher than the amount that users at $Q_1$ or $Q_2$ would pay for its use, the price would be too high to permit use.

This example helps provide further insight into problems of compensation for damage to the environment. The demand curve in figure 7–1 is what potential users would pay to use an uncluttered bridge. However, if congestion develops as described above, the value to users at $Q_3$ and beyond would be zero. Note, this is equally true in the case of land. If the high bidder by definition gets the land, subsequent bidders cannot use it. They would like to, and they would pay a given price if they could. The price we show is the one they would pay if no one else interfered with their preferred use. But that is not possible if high bidders get the bridge and the land. The amounts nonreceivers would be willing to pay would decrease with congestion and occupation, becoming zero at $Q_3$ in the case of the bridge. Yet, whenever measures are made of the damage done from environmental pollution, they are based on the difference between the existing condition and a state of perfection. A measure of damage from smoke is based on a comparison with conditions that would exist if there were no smoke at all.[1] Similarly, users of a highway or bridge would measure their loss against conditions where there was no congestion whatever. We shall cast the problem in another context in the next section of this chapter.

In addition, there is the long-run problem, also discussed by Musgrave (1959), of collecting enough revenue to finance construction of the bridge. If nonusers, rather than owners, are paid, no bridges or highways are likely to be built.

Not all writers argue for compensation to the damaged party. For example, Baumol says (1972: 311–312):

It is important to observe that, *the solution calls for neither taxes on $x_3$, the neighboring laundry output, nor compensation to that industry for the damage it suffers*.

One way to look at the reason is that our model (and the pollution model in general) refers to the important case of *public* externalities. The laundry whose output is damaged by smoky air does not, by an increase in its own output, make the air cleaner or dirtier for others. Hence, the appropriate price (compensation) to a user of a public good (victim of a public externality) is zero except, of course, for lump sum payments.

This proposed solution to the problem is based on another confusion that sometimes appears in the literature—the difference between public and private goods. An understanding of this difference helps to explain why compensation to damaged parties can produce an economic optimum and equilibrium in some cases, where only two parties and two uses are involved, and not in others, such as the bridge and land examples and possibly in Baumol's laundry and smoky industry example.

### Public Good and Private Good Aspects of Environmental Quality

The problems of water quality and air quality are often discussed in the context of a public good. Indeed, when economists look for examples to illustrate the problems associated with public goods, they frequently turn to air and water quality. Whatever water quality flows by A's house is the same that B receives if B's house is next door or on the opposite bank, or it is at least a function of A's water quality if B lives downstream. In analyzing the problems of demand for water quality by downstream users, it often becomes appropriate to treat water quality as a public good, with a vertical summation of individual demand curves to determine the total demand by particular users of water quality.

However, there is also a private-good aspect to water quality, from which the most important questions arise and the most frequent oversight occurs. Water quality may be sought by different users whose uses may be mutually exclusive to some degree. The same situation characterizes private goods. If there is one pair of shoes and A gets them, B does not. If there is a given water quality and A gets to use it in a way that precludes B's preferred manner of use, it is a private good. If an industry discharges waste into a river and downstream swimmers cannot then use the water, water quality is a private good.

The case in which use by swimmers is a private good has been more difficult to see. So long as crowding is no problem, swimmers who live either upstream or downstream from the mill do not affect the use of the water by other swimmers, which means that their collective use of water may be analyzed as a public good. In fact, if swimmers live upstream from the mill, they do not affect the use of the stream by the mill and the use of the stream may be analyzed as though it were a public good for all parties. But if the swimmers live downstream from the mill, the stream is a private good to be allocated among those upstream and downstream who want to use it. An accepted economic solution for the use of a scarce private good is bidding by all competing parties desiring the use of the good or service, or, lacking private ownership of the good, at least levying a charge, tax, penalty, or fee on all users of the good or service. But because the problem of pollution usually has not been seen as one involving the allocation of a private good, the solution commonly proposed has been to charge only one party, instead of charging both or all competing parties who use the good as is done when allocating a private good.

Not only has this failure to see the private-good aspects of the use of environmental quality created problems in the pricing of the products; it has also produced confusion when applied to compensating certain parties involved in problems of pollution.

Baumol (1972) rules out compensation to the owner of the laundry whose output is affected by smoke from the neighboring industry's plant, because the laundry's use of air is the use of a public good. There is an element of truth here, when one laundry's use is related to use by other laundries. But with regard to neighboring smoke-producing firms, the use by this particular industry, to the extent that it gets its way in choosing and securing the air quality it prefers, is that of a private good. Note that in Baumol's solution, the laundry does get a reduction in smoke, at the expense of the smoke-producing industry. In this use of the environment as a private good, present-day problems arise.

There is another feature that must be considered to explain why compensation seems to provide an optimum in our earlier upstream-downstream case and in many environmental quality cases, but not in the highway and land-use cases discussed above. In the usual water quality example, there is one firm upstream and there are many users downstream. Those who are downstream use water quality for a single purpose as a public good, so that the vertical sum of their demand curves shows the value of water quality in one particular use. Analytically, they may be treated collectively as one user. With one user upstream and effectively one user downstream, figure 3–1 depicts the situation.

As a solution to allocating some private good where only the two uses are possible, the result is optimal and may be obtained either by the usual process of having each party pay for the good or by having either party pay for the good while the other is compensated, abstracting from income effects. If payment and compensation are employed, the former will equal the latter in marginal terms at the optimum and the sum offered will always exceed the sum asked as compensation. If the water quality in a stream were privately owned, the cost to the owner when those upstream (downstream) used the water quality would be the loss of income from those downstream (upstream), the only other water user(s). Thus, a system that charged one party and compensated the other would yield the same result as one that charged either party and compensated the owner, provided that the one not charged could not negotiate with the other party for further adjustment.

It was an easy, but improper, step to extend this analysis to the case of Musgrave's bridge. Here was a public good, in that many persons could use the bridge (up to the point of congestion) without any adverse effect on other users. There was also the private-good limitation after congestion, when one more user would reduce the advantage to all others on the bridge and perhaps preclude the use by still others who were not yet users but would like to be. However, there is a problem that is more obvious in

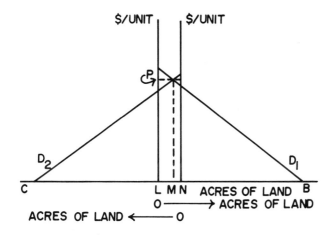

**Figure 7–2.** Allocation of a Given Amount of a Good (Land) among Many Mutually Exclusive Users

the case of the bridge but that is potentially present in all cases of environmental pollution. Often there are many potential users who wish to use the environment in a multitude of mutually exclusive ways, and the requirement of compensation to these parties for the use of what is basically a private good may preclude its use.

In the case of land, the problem becomes apparent quickly. If compensation were required in the amount bid by each other user, representing the damage he suffers by being deprived of using the land as he would like, the land would never be used.

The case may be seen in figure 7–2, which is a variation of figure 3–3(c). Assume that the 100 demanders of land each want one acre of land to be used in a way that precludes use by any other party. The total demand curve may be broken down into two demand curves. All persons who bid even amounts may be added together (horizontally) in one demand curve, and all who bid odd amounts may be treated similarly in another curve.[2] The first curve can slope downward to the right from one axis and the other downward to the left from the opposite axis. If only two acres are available, the two vertical axes will be separated by a distance representing that quantity of land. The quantity of land will be allocated so that $LM$ goes to the highest bidders represented by $D_1$, and $NM$ goes to the bidders represented by this portion of $D_2$. Each bidder pays a price of $LP$, if one price is charged all purchasers. Those demanders represented by $MB$ on demand curve $D_1$ and by $MC$ on demand curve $D_2$ are damaged by being deprived of their preferred use of the land. However, there is seldom, if ever, a plea by those who are outbid that they should be compensated because of this damage. Rather, payment goes to the owner of the land. Further, in the case shown in figure 7–1, the amount paid could not compensate the damaged parties for the loss they suffer, as measured by the amount they would be willing to pay for the land.[3]

The problem under a system of compensation arises because 100 persons have each been given ownership of one acre of land when only two acres are available. Further, ownership carries the right only to sell rights to the land. Ownership does not carry the right to use the land, unless it is a sole ownership. Thus, if one of the parties wants to use it, he has to pay the other potential users (owners), or if our previously successful bidder becomes an unsuccessful bidder, he is now entitled to compensation. If we require that all parties who are damaged be compensated, we may be distributing rights to more assets than exist. In such a case, everybody's asset will become nobody's asset, and its use will go by default to those who prefer it in its natural state, or are strongest, first in line, favored by gravity or wind, or capable of generating the loudest or most persuasive arguments. Thus, use of city streets may go to the group that can control the area, use of a congested highway goes to those who are first to enter it; the mill upstream may use the river while swimmers downstream cannot; or Sierra Club members may persuade Congress that their intended use of the Alaskan tundra is superior to that sought by others and so benefit from a legal right to enjoy the environment in their preferred way.

Musgrave's example of the bridge is similar. If there are 5,000 potential users and each would pay a different sum for its use, starting with $50.00 as the highest bid, with subsequent bids declining at the rate of one cent per bid, the results could be the same as in the land case just cited. Granted, because of the public-good nature of limited use, perhaps 100 drivers could use the bridge at its optimum output, but the sum of their payments would not be adequate compensation for those denied use. Other examples could be formulated where the amount paid would cover the amount due as compensation; but this would still leave the problem of financing the bridge, as it would for any other man-made assets.

When there are only two potential uses or users of a given asset, it necessarily follows that the amount paid by the higher bidder must be sufficient to compensate the lower bidder, if that should be necessary. When there are more than two potential uses, this sufficiency no longer necessarily holds. The point appears to have been overlooked in previous discussions of environmental quality because only two uses of any given part of the environment usually have been hypothesized. Further, because of the public-good nature of one of the uses usually described, it appeared that the two-party solution applied even when there were many users. In cases limited to two competitive bidders, payment by one party and compensation to the other party produced both an optimum and an equilibrium.

But this overly-simplified model may fail when there are more than two mutually exclusive uses to which some environmental asset could be put. The example of the stream could be broadened to involve the upstream mill and downstream swimmers, water skiers, fishermen, and picnickers. It is possible, perhaps probable, that on a small body of water, each use would by itself preclude any of the other uses. A rule that the mill must compensate those damaged by its wastes could prove prohibitive, when a limited discharge may be the optimum use of the stream. Swimming, fishing, water skiing, and picnicking may also be as mutually exclusive as competing uses of a given plot of land. The requirement that any downstream user compensate all other parties whose use is forgone may be equally prohibitive. Even Baumol's example of air quality may involve many uses of a private good. If the laundry wanted air quality that included no smoke and no moisture, a spinning mill wanted air with no smoke but much moisture, and a steel plant wanted air with much smoke and moisture from emissions, any of these users would have to compensate two other parties who were damaged by his use. There could be three firms, or several firms in each of the three industries; but each of the three would be a user of what is, among the three, a private good. Within any of the industries, however, the air may be a public good.

**Technological and Pecuniary Externalities**

Social costs are usually discussed under the heading of externalities. The discussion of external effects, as Mishan (1965) observes, has done much to clarify the issues. But the discussion has not been entirely beneficial, because the term has become too general and, as a consequence, ambiguous. A failure to view specific problems, such as the classic case of smoke in the air, as a problem of asset use has led writers to a restricted view of external effects that has made more difficult the formulation of a general and useful definition of externalities.

The distinction between pecuniary and technological externalities was proposed by Viner (1931). Pecuniary external diseconomies occurred when one industry expanded its use of factors and materials so that other industries had to pay higher prices for them. Technological external diseconomies were reflected in increasing technical coefficients of production as the output of an industry expanded. Thus, pecuniary external diseconomies were reflected in changes in prices, while technological external diseconomies were reflected in required changes in the quality of factor inputs. However, these are just two of the dimensions of a three-dimensional supply plane (or multidimensional supply function) for a given asset. The variables to be considered are quantity, quality, and price. As is well-known, if a new plant moves into town and hires workers, the price of labor of a given quality rises. If, instead, a plant moves into town and puts its wastes into a stream, the price of water of a given quality rises. There is little, if any, difference in the two cases except that in the former, the effects normally are expressed in market prices, while in the latter, the effects normally are expressed in quality levels (Bator 1958: 358). An obvious way to analyze the problems of technological externalities, then, is to assume a policy of pricing analogous to that used when pecuniary externalities arise, but the use of this procedure has been limited.

In a footnote, Baumol (1972) seems to treat pecuniary externalities as a case apart from the one he cites in which the smoky factory hurts the laundries. He quotes a proposal by Coase (1960) to tax (charge) residents who move near a mill because their presence will necessitate greater expenditures for purity by the mill.

Like a price change, the variation in taxes constitutes a pecuniary externality [which does not lead to resource misallocation]. Both have real consequences but they are merely 'movements along' the production and utility functions, i.e., any given vector of inputs will be able to produce the same outputs as before the change in tax rates, and any vector of output levels will still be able to yield the same utility levels (p. 312).

But either production and utility functions are highly generalized and people move along them in many directions based on varying prices and qualities, or the functions are highly specific and people move along them based on specific prices and qualities.

Baumol sees that a factory can impose externalities on neighboring inhabitants, but he fails to see that neighboring inhabitants can impose externalities on the factory. He argues for pricing the former externality, by a tax, so it will be reflected in market decisions. If the problem had involved a plot of land that was desired by a factory and neighboring residents, it is doubtful that Baumol would have proposed a similar policy. Such a policy would require that the government or some other agency determine how much each party values additional units of land, choose the price and allocation that yields the optimum, sell to the factory the land it gets, and give to the residents the land they get. Again, this solution could be an optimum; but because of the failure to price land for residential use, it would not be an equilibrium.

This tendency to overlook the role of mispricing an asset in the creation of externalities has led other writers to concentrate instead on the effect of externalities on the production function. Thus, Mishan (1965) makes a distinction between (1) external effects internal to the industry (such as occur when many fishermen who are members of the same industry overfish a site that is not priced) and (2) external effects external to the industry (such as occur when the steel industry produces smoke that damages firms in the laundry industry). He then notes (pp. 6–7):

> In regard to (1), if some factor used in industry A is, say, undervalued in consequence of the industry's internal organization, there does not occur any alteration in its production function. . . . In regard to (2), however, inasmuch as industry A generates external economies *external* to itself, it alters the production function of some other industry, say B, and this *does* affect the technological possibilities of production.

In each case, however, an asset is underpriced and overused: in the first case, to the disadvantage of other or all firms in industry A, where economic efficiency would be improved if some of the firms in industry A were forced to do without the use or do with less use of the favored factor; and in the second case, to the disadvantage of some or all firms in industry B, where, again, economic efficiency would be improved if some of the firms in industry A were forced to do without the use or do with less use of the favored factor. The discussion of shifting production functions appears to be a digression that obfuscates rather than clarifies, the issue at hand.

Continuing his discussion of the second case above, Mishan notes (p. 7):

> If B were a perfectly discriminating monopolist and, therefore, already producing an optimal output, the resulting effect on B's cost curve indicates the new optimal output which B will now produce. Industry A's output alone will require correction according as the external economy, or diseconomy, generated varies with its own production. Once effected and a Pareto optimum realized, however, the presence of this external effect is not—in contrast to (1) above—completely eradicated.

If the problem is due to mispricing the scarce asset, the introduction of a price will require that both industries, A and B, change their outputs, unless either is faced with a perfectly inelastic demand or supply curve. External effects will be eradicated in both cases or in neither case, depending on how one defines external effects. With proper pricing, all parties will have to adjust their use of the asset and their outputs, and no unpriced (external) effects from this source will remain.

A third, but minor, difficulty arises perhaps from Mishan's point that "external effects may be said to arise when relevant effects on production or welfare go wholly or partially unpriced" (p. 6). By concentrating on effects instead of assets, people have developed a tendency to overlook a part of the problem. The effect of smoke on neighboring industries is obvious, but the effect of neighboring industries on the smoke-producing industry is far from obvious. If the definition were changed to read, "External effects may be said to arise when relevant assets used in production or for welfare go wholly or partially unpriced," the problem would be more easily observed and solved.

## Summary and Conclusions

There are at least four concepts that have been misinterpreted and have thus contributed to the formulation of inefficient solutions. First, pollution has been seen as a physical phenomenon instead of an economic one. Though uncleanliness is obviously damaging to those who desire cleanliness, it has not been obvious, or easy to understand, that cleanliness may be damaging to those who benefit from uncleanliness. Even where the latter relationship has been acknowledged, the usual policy has been to forbid arbitrarily most uncleanliness. Indeed, the concept of cleanliness is itself indefinite, sometimes meaning a change from the natural state of the environment and sometimes not.

Second, a concern for fairness and equity has led to the idea that people damaged should be compensated, but this overlooks the fact that in cases involving conflict, each and every party is damaged unless he gets his way entirely and has no restrictions placed on his preferred actions. If compensation is paid, in effect property rights have been assigned to certain persons. There are two problems with this procedure, however. In many, if not most, cases, more rights will be assigned than there is property represented by these rights. Further, by concentrating on compensation, writers have stressed fairness without observing its effects on efficiency or the cost of such fairness. In addition, fairness is none too firm a reed on which to rely in allocating property rights.

As a third area of difficulty, the stress on the public-good nature of most assets that create social benefits and costs has led to an oversight about their private-good aspects. Many writers have observed that one swimmer does not reduce the ability of other swimmers to enjoy the water, and these writers have concentrated on finding the optimum level of use by swimmers, without considering the effect that cleaner water for swimmers may have on other users of water. There is a place for the analysis of the use of an asset as a public good, but unless there are no competing uses of the asset, allocative efficiency requires that we also analyze the private-good aspects of these assets.

Finally, the stress on technological externalities has clouded the fact that they are no different analytically from pecuniary externalities. The difference lies in ready markets in which price changes are observed in the latter case, and the lack of such markets in the former case. But the lack does not mean that price effects cannot be considered or that they will not be realized by bargaining between the parties concerned, or by a charge by a third-party owner of the asset. The stress on the technological externality effects created has hidden the true nature of the problem, the existence of some unpriced asset whose use is desired by more than one user or group of users. If the problem is viewed not as smoke damaging the laundry, but as two users desiring to use air quality, solutions become simpler.

Economists frequently note that economic theory can contribute to the solution of problems that do not involve markets but that do involve scarcity. The problem of environmental quality and its use is one such problem. Our principal concern should be to cast the problem in the existing economic framework and not to treat the problem as a special case involving pollution, compensation, public goods, and externalities. With Occam's Razor we can cut away much of the haze and get down to the simple concept of a scarce asset. Then the problem, if not the air, will be much clearer.

## Notes

1. For an example of such a statement of loss, see an analysis by the President's Council on Environmental Quality of the loss from the pollution of the Delaware River Estuary in James Hite (1972: 104–105). There is far more to pricing under conditions of congestion than is indicated here. Some of the most important aspects are covered in the classic article by Frank Knight (1924). William Vickrey (1967) has also discussed optimization under conditions of congestion.

2. While the prices described are discrete, the curves have been drawn in the more general continuous form.

3. Coase has pointed out that the amount that a user would be willing to pay for a resource does not necessarily represent the loss he would experience if he did not get it. There may be alternatives that are only slightly less desirable. Thus, if a person would pay $5,000 for a plot of land but he is outbid for this plot and must of necessity turn to a less desirable one, his loss is equal to the reduction in his welfare from taking the less desirable plot less any saving in cost he enjoys. The point is equally applicable to persons who claim they suffer losses from environmental pollution equal to what they would pay for higher environmental quality.

# 8 Other Lessons

The theory of environmental use described above applies to more than just the environment. Because it is drawn from traditional economic theory, we should expect it to have broad application. But as environmental use has not been viewed traditionally as a problem involving scarce resources, it is not unusual that many of our other problems have not been viewed in that light, either. As a general principle, whenever two or more people disagree on some given course of action, the problem can be analyzed as one involving scarce resources. At the bottom lies a disputed property right.

We offer three present-day problem areas that can be clarified using the approach we have described. Stated succinctly, they involve abortion, gas-guzzlers, and economic welfare. More generally, they involve, an application of economic theory to a question that does not appear to be economic; an application of the current special environmental theory to a question that is clearly an economic problem; and a misconception of cost that is easily seen when economic theory is used.

## Extending Economic Theory

Many states are in the process of changing from laws that forbid abortion to laws that permit abortion. Regardless of whether the laws ultimately permit or prohibit abortion, large numbers of people will remain dissatisfied. To be or not to be is not exactly the question. How many should be or not be is a better question. If we view the problem in the light of the now-familiar figure 3–1, we may improve our understanding, and perhaps even our social well-being. Consider figure 8–1, whose marginal benefit curve measures the intensity of desire to have abortions, over and above the normal medical costs, sacrifice of income, and pain and suffering associated with an abortion. The measurement is in dollar terms for marginal units of the operation. Some women want badly to have the operation; others are less enthusiastic; at the zero price level, some beneficiaries are indifferent and should not be described as beneficiaries.

The marginal dislike curve measures the intensity of feeling of those who oppose abortion. If those feelings are depicted accurately, these people recognize the need for some abortions and do not find these first

119

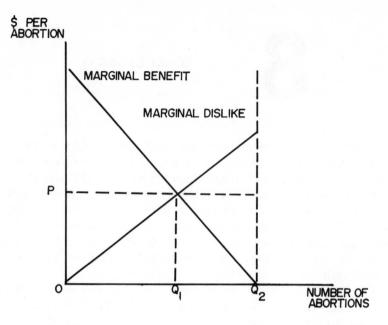

**Figure 8–1.** Marginal Values Associated with Abortion

few so distressing. As the number of abortions increases, however, so does the distress per abortion. Anti-abortionists deal with a public good. Their demand curves are summed vertically. There can also be a public good element in the demand for abortion.

The conclusion is that there is an optimum number of abortions, just as there is an optimum level of pollution, or purity. What is missing to make this operative is a system of establishing property rights to this operation. These may be given to each woman, so that those who oppose abortion would have to pay pregnant women who reportedly forwent abortion. This presents a good opportunity for couples desiring children to collect from anti-abortionists. Alternatively, the rights could be given to the Anti-Abortion League, and rights sold to those who badly want the operation. It would be interesting to see what sum would salve the moral conscience. Further, many citizens could suddenly develop a feeling in opposition to abortion and so benefit from the payments made. A third solution is perhaps preferable whenever two parties disagree over some question where property rights have not yet been clearly established: sell the rights to a third party, allowing him to maximize his income subject to selling at the same price per unit to all buyers.

In this way, the owner of the rights can estimate the conditions in figure 8–1 and try to determine the equilibrium price. Each person buying an abortion must pay an abortion fee in addition to medical and other fees. The Anti-Abortion League would collect from interested parties or members and buy a reduction in abortions shown by $Q_2 - Q_1$. All the conditions and advantages of a price system cited above would apply here. People who want abortions can have them, provided they pay all the costs associated with the operation, and part of these costs is the misery imposed on anti-abortionists. Those who suffer are not compensated, for the reasons given above: everybody suffers. Instead, the owner of the rights is compensated. Those who oppose abortion could eliminate abortions entirely, if their intensity of feeling were so strong as to lead to payments that were greater at the margin than the price anyone would pay to have an abortion. Once again, payment (and happiness) would be greater than the amount anyone else would pay (and by not paying, suffer). Although the operation of such a system is unlikely under present mores, tastes, and concepts of rights, at least the analysis helps us to see that neither the absolute prohibition nor the unlimited permission of abortions is an optimal solution.

A less serious but more tractable example concerns the question that arose at several Southern schools in the late 1960s. Should *Dixie* be played at football games? It had long been a tradition to play the song, and many people wanted to continue hearing it frequently. Others disliked hearing the song and wanted it banned. In most cases, people resolved the question by voting. To play or not to play, that was the question. To pay or not to pay would have been a better question. A variation of figure 3–1 could be used again, with one group paying to hear *Dixie* played the optimum number of times and the other group paying to avoid hearing *Dixie* the optimum number of times. The frustration factor should be reduced, as people would know why they heard the song only, say, 2.7 times, or why they had to hear it 2.7 times.

The broad application of these principles may perhaps be realized if the reader uses them to analyze problems of population (in the world, New York City, or Oregon), freedom of speech, billboards, and trucks on highways. Even income distribution should be included if some people are jealous of higher incomes received by others and thereby suffer negative externalities. But they will have to pay a price to see Peter robbed and Paul paid.

## Pollution Economics Improperly Applied

We have seen that viewing environmental use as a problem of pollution will probably give us a limited view and a nonoptimum solution. However, pollution is still the accepted view and it crops up in places where it formerly did not. More technically, we have turned pecuniary externalities into technological externalities. For example, it is obvious that when a luxury car is driven a hundred miles, it uses more gasoline than does an economy car. This means that the price of gasoline rises or that less gas is available for the economy car. In turn, this makes the environment (broadly defined) of the owner of the small car less pleasant, or polluted. While the market has been busy allocating gasoline to those who treasure it highest, relative to other goods, those who treasure it less highly or have limited incomes and large desires find they have been hurt, compared to what would happen in a world of plenty. To reduce this damage, Congress has decreed that those who want and are willing and able to pay for gasoline to propel large cars will have to buy smaller cars so that people who prefer smaller cars may be able to get more of the limited resource. Although the proposal is supported on the basis of conserving energy, the burden falls on only one group of users, with higher relative desires but fewer votes than the other group.

Similar concern has been voiced for people who are unable to afford cereal grains because other people prefer to have their grain converted into meat. Proposals have been offered to raise the price (the polluter must pay) or to reduce (regulations will restrict) the quantity of meat produced. In a related case, exports of grain to foreign countries have been forbidden, so domestic consumers of grain will suffer less "pollution" of their total environment because of higher grain prices. It has been pointed out above that when resources are scarce, every user suffers to the extent that he does not receive all he wishes at a zero price. The market is designed to minimize this suffering, or, stated more positively, to maximize satisfaction. By viewing and treating pecuniary externalities as pollution to be avoided, the prospects for inefficiency are limitless.

**The Cost of Pollution**

William Nordhaus and James Tobin (1972) have attempted to obtain a Measure of Economic Welfare (MEW), and Samuelson (1973) has renamed this Net Economic Welfare (NEW) to reflect more accurately some of the things done in measuring the value of output. Specifically, both subtract disamenities and externalities. "The disamenities of urban life come to mind: pollution, litter, congestion, noise, insecurity, buildings and advertisements offensive to taste, etc. Failure to allow for these negative consumption items overstates not only the level but very possibly the growth of consumption" (Nordhaus and Tobin 1972: 49–50).

But these disamenities are little different from the ones we all suffer every day from failing to get all of the other goods that we want. If failing to get clean air is a disamenity, so is failing to get good land. Failing to enjoy limitless leisure while working at some job is also a disamenity. Figure 3–1 should help us to see the point.

If any adjustment is required, it should be made to increase the value of gross national product by the cost of the unpriced factors used in producing it. If the polluter pays, that payment raises the cost, and value, of the good. Samuelson points out that we should do this and also subtract a like amount for the disamenity, which would leave the original GNP value unchanged. We argue that there is no reason to subtract disamenities in the case of air, water, or noise, unless they are subtracted in cases of land, labor, and capital. When resources are scarce, everybody suffers. But everybody also gains. More precisely, those who want, but not badly enough to pay, suffer. And those who want and will pay, gain. As Figure 3–1 reminds us, these are often the same people. GNP may not be a perfect measure, but it will be a better measure without subtractions for disamenities and externalities.

## Conclusion

Robert Nozick (1974) has commented that Marxism is an exploitation of the people based on their ignorance of economics. Perhaps the ideas presented above will help to remind us of what we know in economics and to prevent exploitation of people and their environment.

# Appendix A
## A Fable[a]

Once upon a time there was an uncrowded nation which was richly endowed with natural resources in the form of mineral deposits, forests, and water resources embodied in streams, lakes and groundwater aquifers, and wherein there existed great diversity of climate, soil and topography. In this nation there were two great regions, Metropolitania and Ruritania.

For many years, Metropolitania had grown in population, wealth and income until its industries were large and diversified, its towns and cities numerous, and most of its people prosperous. Even Metropolitania's poor were well off by the standards of many other nations and regions, including Ruritania, and as time passed the poor and the disadvantaged from Ruritania moved in large numbers to the cities of Metropolitania in search of better jobs, higher incomes, greater economic security, and better community facilities and services. Much of the labor force employed in Metropolitania's industrial, commercial and financial enterprises had migrated from Ruritania after being reared and educated there, and much of the fuel and raw materials used in Metropolitania's electric power stations and manufacturing plants were also imported from Ruritania.

As time went by the streams and lakes of Metropolitania gradually became more and more polluted by sewage and other waste materials from its cities, industrial plants and public facilities; the air above and near urban-industrial areas became murky, foulsmelling and unhealthful; garbage and trash disposal became more costly but less effective, and the costs of local governments rose because of problems growing out of the inability of migrants from Ruritania and similar regions to cope with the opportunities and problems they encountered in an urban-industrial environment. As the costs of government gradually rose and the quality of the environment gradually declined in Metropolitania, its superiority over Ruritania for the location of new and expanded industrial facilities and as a place to rear and educate a family was gradually reduced. Consequently, there developed a growing tendency for Metropolitanian industrial firms to build branch plants in Ruritania where the air was clean, the water pure, good industrial sites plentiful, taxes low, and the general public and public officials grateful for the economic opportunities and benefits provided by new industries. As the economic base of Ruritania

[a] Reprinted with the kind permission of Professor James M. Stepp and the American Enterprise Institute for Public Policy Research from James C. Hite, Hugh H. Macaulay, James M. Stepp, and Bruce Yandle, Jr., *The Economics of Environmental Quality*, (Washington: American Enterprise Institute for Public Policy Research, 1972), pp. 53–61.

expanded and diversified, its towns and cities grew, its public facilities improved, and its public officials became more aggressive and more successful in competing with Metropolitania for the location of new industrial and commercial facilities.

The serious decline in the quality of the physical environment of the people of Metropolitania was so gradual that for many years it was noticed only by a few observant and knowledgeable persons, but there came a day when the eyes of the people were opened and they saw and comprehended the degradation and ugliness of their environment. There was a great outcry against those who were polluting the land, waters and air of Metropolitania, and a popular demand arose that pollution be prohibited by law and violators punished. The political leaders of Metropolitania conferred among themselves and came to this conclusion:

Our people demand that their environment be cleaned up and that steps be taken to preserve its quality, and if we are to continue to be leaders among our people we must find an effective and satisfactory way to bring this about. But if we increase taxes to build and operate waste disposal and waste treatment facilities our people will be angry despite the fact that they are rich and prosperous; and if we compel industrial and commercial establishments, at their own expense, to reduce drastically their pollution of the environment, they may build more and more of their new and expanded facilities in Ruritania where the land is uncrowded, clean water plentiful, labor abundant, and the people more interested in improving their economic status than in preserving the purity of their streams and atmosphere. To place a heavy burden upon ourselves, on the other hand, could cause the economic growth of Metropolitania to slow down and the prosperity of our people to decline. We must therefore devise a plan of pollution control which will not impair our ability to compete with Ruritania. Let us hire a consultant to develop such a plan.

After many weeks of study and many postponements the consultant, a loyal resident of Metropolitania, made the following report:

My studies have revealed that it is possible to control pollution in a way that will not only preserve Metropolitania's ability to compete with Ruritania for new industrial facilities, but will actually reverse the tendency of many corporations to build their new plants in Ruritania. This industrial migration has already harmed the residents of Metropolitania, especially those Metropolitanians who spend their vacations in Ruritania and enjoy the natural beauty of the countryside and the quaint customs of the people. In many Ruritanian communities industrial development has already spoiled the natural environment, changed the customs and life style of the people, and raised their incomes so much that they are unwilling to accept seasonal jobs at reasonable wage rates. The best place for a person with a good job to spend his vacation is an uncrowded area where the water is clean, the air pure, and the local residents so poor that they are willing to accept seasonal jobs with irregular hours and low pay, and these conditions are already ceasing to exist in parts of Ruritania.

The first thing we should do is to persuade our national government to proclaim that environmental pollution is a *national* problem rather than a local or regional problem, and that there is a *national* interest in the quality of the water in every segment of every stream, in the quality of the air over every square mile of land and water, and in the amount of junk and trash on every vacant lot in the nation. It would follow, of course, that since the polluted land, water and air of Metropolitania constitute a *national* problem, the taxpayers of the whole nation, including the residents and the industrial enterprises of Ruritania, should bear the costs of cleaning them up and keeping them clean. It would also follow that the clean environment of (most of) Ruritania is a *national* asset whose quality should be preserved and enhanced for the benefit of *all* of the people rather than exploited and degraded for the economic benefit of selfish interests.

The political leaders of our nation should be persuaded to enact laws and promulgate regulations granting financial assistance for water and air pollution control by existing industrial plants. Such laws and regulations would benefit the residents of Metropolitania in several ways. First, old industrial plants (of which Ruritania has comparatively few and Metropolitania many) will encounter more technical problems and higher costs of pollution control than new and modern plants (which are already fairly numerous in parts of Ruritania and threaten to become more and more numerous there); hence our industries would receive more financial aid per worker and per unit of output than theirs. Second, the increased taxes required for the national government to provide such financial assistance to industrial plants would, or could be made to, fall more heavily upon the economy of Ruritania than of Metropolitania. To the extent that new industrial plants are more profitable than old ones, average corporation income tax collections should be relatively higher from Ruritania than from Metropolitania. Also, since industry provides a much smaller percentage of the jobs and income in Ruritania than in Metropolitania, increases in the personal income and sales taxes paid to the national government will be much less per dollar of pollution control subsidy here than there. Third, if the financial aid were made to include the usual provision that contractors building such facilities pay "prevailing" wage rates as determined by the appropriate agency of the national government, the costs of such facilities in a labor-plentiful, low-wage region such as Ruritania might in many cases be raised so high above those of typical low bidders that they would choose to forego financial aid to avoid what would appear to them to be artificially high labor costs. Choosing to forego financial aid would not, of course, excuse Ruritanian firms or residents of Ruritanian cities from paying their proportionate shares of the taxes imposed to support the program.

It is clear, therefore, that a national government program of financial assistance to industries and municipalities for pollution control can be used to restore and protect the competitive position of Metropolitania relative to Ruritania. Such restoration and protection can be promoted even more effectively, and more directly, by establishing the proper kinds of rules, regulations, restrictions, and standards governing the use of the natural environment for waste disposal. For example, all persons, firms, and agencies could be prohibited from discharging any liquid wastes into any stream or other body of water without treating such wastes sufficiently to prevent any reduction in the quality of the receiving body of water. This straight-forward and plausible-sounding approach to prevent any further deterioration of water quality in the nation would have the effect of converting the numerous clean streams of Ruritania from assets into liabilities for purposes of economic growth and development, and the numerous polluted streams of Metropolitania from liabilities into assets. An industrial plant or municipality discharging wastewater into a clean stream would be required to remove

100 percent of the waste materials from its effluent, which would be very costly, while one discharging its wastewater into a polluted stream could comply with the law by removing a much lower percentage of the waste materials at a much lower cost per gallon of wastewater. Requiring that those discharging wastewater treat it sufficiently to *raise* the quality of the receiving waters would affect interregional economic competition between Metropolitania and Ruritania in essentially the same way, and for the same reasons.

One might question the political feasibility of a pollution control program which would benefit a wealthy region such as Metropolitania at the expense of a poor region such as Ruritania if the two regions were fairly equal in population and political power or if a substantial percentage of Metropolitanians were strongly concerned with alleviating poverty, ignorance and disease in Ruritania. Fortunately, neither of these conditions exists. A large majority of the citizens of the nation live in Metropolitania, and most of them can be counted on to support a policy and program which will confer both economic and environmental benefits upon them and their neighbors. The liberals and intellectuals who might ordinarily be expected to defend the economic interests of the poor and disadvantaged residents of Ruritania are so alarmed over environmental pollution and so concerned with environmental protection and improvement that most of them will pay little or no attention to adverse economic side effects of pollution control, especially the side effects which occur in some other region.

Measures such as those I have described should have an excellent chance of being enacted by the national legislature if influential representatives from Metropolitania can be persuaded to sponsor the appropriate legislation.

Again the political leaders of Metropolitania conferred among themselves. They agreed that the consultant's analysis was sound and his plan good. The plan was enacted into law; a bureaucracy for enforcing it was created; detailed guidelines, rules and regulations for compliance and enforcement were promulgated; and there was assembled an energetic and ambitious enforcement staff dedicated to the principle that the quality of the environment is too important to allow its protection to be compromised for the economic benefit of any group or area.

As the years went by the predictions of the consultant were fulfilled. The trend toward locating new wet-process manufacturing plants in the humid parts of Ruritania was abruptly reversed shortly after the new law went into effect. Since average wage rates in these industries were considerably above the average for all industries, the adverse effect upon the growth of wage and salary income in Ruratania was even greater than the adverse effect upon the growth of employment. Since most wet-process industries are capital-intensive, the adverse effect upon the growth of property and income tax revenues was even greater than the effect on wage and salary incomes. Opposite effects were simultaneously occurring to Metropolitania as more and more new wet-process manufacturing plants were being built where the firms could avoid the rigorous and costly wastewater treatment requirements of Ruritania and at the same time receive public praise and also public subsidies for treating their effluent sufficiently to raise somewhat the quality of Metropolitania's polluted streams and lakes. Thus it was that the wealth, income, public services and economic security of the residents of Metropolitania began to grow as the quality of their environment gradually improved.

In Ruritania the quality of the natural environment also gradually improved as numerous manufacturing plants and municipalities were compelled to reduce to almost zero the amounts of waste materials discharged. But, since some of the older industrial plants could achieve such a level of waste reduction only by closing down, employment in wet-process industries decreased and tax revenues from such industries decreased by even greater percentages. At the same time the residents of Ruritania demanded that their local governments provide expanding and improving transportation, education, health, recreation and welfare services similar to those reported by news media as being provided in Metropolitania. The small size of the tax base made it impossible, however, to provide local government services that were even close to the quantity and quality of those provided in Metropolitania, but the gallant effort of Ruritania's elected officials to do so caused tax rates to rise to levels substantially above those applied to the much larger per-capita tax base in Metropolitania. These high taxes on the property, sales and income of business enterprises severely impaired the ability of Ruritanian manufacturing plants to compete with Metropolitanian plants in national and foreign markets even when they had no waste disposal problems; hence there developed a tendency for all kinds of industrial firms to avoid locating new plants in Ruritania unless raw materials or markets provided some very special reasons for doing so.

As old Ruritanian plants were closed and replaced by new plants located in Metropolitania, the Ruritanian tax base shrank and the quality of local public services deteriorated to such an extent that officials of large firms became less and less willing to live and rear their children in Ruritania despite the fact that the natural environment was far cleaner than that of Metropolitania. They preferred to live and work in Metropolitania and visit Ruritania only for short vacations.

As large industrial firms closed down their Ruritanian branch plants, their best workers were offered jobs in other plants located in Metropolitania, and the most ambitious and adaptable ones accepted. Also, regardless of whether the employees of a closed plant were offered another job by the same firm, the most able, ambitious and energetic ones migrated and found employment in one of Metropolitania's numerous towns and cities. As the years went by the most able, ambitious and energetic children of the remaining residents of Ruritania also migrated to Metropolitania to take advantage of the superior economic, social and cultural opportunities there, and as the quantity and quality of the labor force declined, the rate of new and replacement investment, the number and quality of job opportunities, and the tax base in Ruritania also continued to decline in what amounted to a vicious circle of poverty, indolence, ignorance, and inadequate public services (by national standards).

The remaining Ruritanians gradually adapted themselves to their economic and social environment by developing customs and attitudes favorable to irregular work habits and unstructured, noncommercialized leisurely activities (often designated by crude and unsympathetic critics as "loafing") and unfavorable to rigorous and sustained efforts to earn income and accumulate wealth. This culture of poverty and indolence, combined with the fact that Ruritania was sparsely populated and had a high-quality natural environment as compared with Metropolitania, caused more and more Metropolitanians to choose to spend their vacations in Ruritania. Commercial vacation and recreation centers and facilities were developed (mostly with Metropolitanian initiative and capital), and these provided seasonal employment for considerable numbers of Ruritanians as well as Metropolitanian college students who needed temporary jobs.

The fees charged by these commercial recreation enterprises were quite high in spite of their low wage scales, not only because of the seasonal variations in the use of facilities, but also because of the high cost of treating their wastes sufficiently to avoid any degradation of the clean natural environment of Ruritania. Coupled with the time and expense involved in traveling from Metropolitanian cities to Ruritanian vacation centers, these high costs restricted the patronage of such centers largely to the more affluent residents of Metropolitania; hence the development of such centers was too limited and scattered to have any appreciable effect upon the economy or the culture of Ruritania.

Thus it came about that a more or less stable relationship was established between affluent but polluted Metropolitania and impoverished but unpolluted Ruritania. Residents of Ruritania who disliked poverty more than pollution were free to move to Metropolitania, and residents of Metropolitania who disliked pollution more than poverty were free to move to Ruritania. There continued to be a certain amount of migration in each direction, but there was no tendency for a strong net movement either way. Eventually the poverty-induced social customs of the Ruritanians came to be recognized by Metropolitanians as a quaint and unique part of the national cultural heritage and thus worthy of protection and preservation.

*Lesson.* This far-fetched and perhaps unrealistic fable conveys an important message about the interrelationships between the environment and the economy of a region despite the fact that it emphasizes only one problem—the possibility that legitimate environmental protection regulations may be deliberately subverted and perverted to serve as instruments of economic protection and/or exploitation by some special group. However, the idea that necessary and desirable measures to protect and promote public health, safety, and welfare may be subverted for the benefit of special-interest groups is neither new, far-fetched, nor unrealistic. On the contrary, such behavior has been fairly common throughout the entire history of the United States, beginning with the disposition of the vast land areas of the public domain and later including freight rate structures, building codes, dairy sanitation regulations, and licensing requirements for numerous trades and professions. The fable is far-fetched and unrealistic only in the single-minded persistence with which it traces out the ultimate economic consequences of a particular stimulus. In the real world of continuous and often offsetting technological, cultural and political changes, ultimate economic consequences are never reached; hence many economic relationships can be identified and clarified only by the oversimplifications embodied in hypothetical models which, like fables, can provide valuable insights into real-world problems.

The extent to which environmental considerations actually do affect various aspects of human behavior and economic competition among people, products and regions is a question of fact rather than of theory.

# Bibliography

Babcock, Richard F. *The Zoning Game: Municipal Practices and Policies*. Madison: The University of Wisconsin Press, 1966.

Barnett, Andy H. "State Tax Incentives for Industrial Water Pollution Abatement." Master's thesis, Clemson University, 1970.

Barnett, Andy H.; Shannon, Russell D, and Yandle, Bruce, Jr. *Establishing Markets for Water Quality*. Water Resources Research Institute, Clemson University, Report No. 44, June 1974.

Barnett, Andy H., and Yandle, Bruce Jr. "Allocating Environmental Resources." *Public Finance*, 28 (1973): 11–19.

Barnett, Jonathan. *Urban Design as Public Policy*. New York: Architectural Record Books, 1974.

Bastiat, Frederic. *Economic Harmonies*. Irvington-on-Hudson, N.Y.: The Foundation for Economic Education, 1964.

Bator, Francis M. "The Anatomy of Market Failure." *Quarterly Journal of Economics*, 72 (August 1958): 351–379.

Baumol, William J. "On Taxation and the Control of Externalities." *The American Economic Review*, 52 (June 1972): 307–322.

Becker, Gary S. "Competition and Democracy." *Journal of Law and Economics*, 1 (October 1958): 105–109.

Boorstin, Daniel J. *The Americans: The Democratic Experience*. New York: Random House, 1973.

Bosselman, Fred P. *Alternatives to Urban Sprawl: Legal Guidelines for Governmental Action*. National Commission on Urban Problems, Research Report No. 15, 1969.

Boulding, Kenneth. "The Economist and the Engineer: Economic Dynamics of Water Resource Development." In *Economics and Public Policy in Water Resource Development*, ed. by Stephan C. Smith and Emery N. Castle, pp. 82–92. Ames, Iowa: Iowa State University Press, 1964.

Boulding, Kenneth E. "The Economics of the Coming Spaceship Earth." In *Environmental Quality in a Growing Economy*, ed. by Henry Jarrett, pp. 3–14. Baltimore: The Johns Hopkins Press, 1966.

Breckenfeld, Gurney. *Columbia and the New Cities*. New York: Ives, Washburn, 1971.

Buchanan, J. M., and Stubblebine, W. C. "Externalities." *Economica*, 29, (November 1962): 371–384.

Burrows, Paul; Rowley, Charles, and Owen, David. "The Economics of Accidental Oil Pollution by Tankers in Coastal Waters." *Journal of Public Economics*, 3 (August 1974): 251–268.

Clawson, Marion. *Suburban Land Conversion in the United States*. Baltimore: The Johns Hopkins Press, 1971

*Clean Air and Water News*. "Charges Press Misled on Facts About Water." 3: 12 (25 March 1971): 177–178.

―――. "Revenue Losses Laid to Delays in Pipeline Approval." 3: 26 (2 July 1971) 397.

*Clean Water Report*. "USC Investigators Report Little Permanent Damage from Santa Barbara Oil Spill." 3: 2 (February 1971): 15.

Coase, Ronald H. "The Problem of Social Cost." *Journal of Law and Economics*, 3 (October 1960): 1–44.

Committee for Economic Development. *More Effective Programs for a Cleaner Environment*. New York: Georgian Press, 1974.

Commons, John R. *Legal Foundations of Capitalism*. Madison: University of Wisconsin Press, 1957.

Costonis, John J. "Development Rights Transfer: Easing the Police-Power-Eminent Domain Deadlock." *Land Use Law and Zoning Digest*, 26: (1974): 6–10.

Dales, John H. "Rights and Economics." In *Perspectives on Property*, ed. by Gene Wunderlich and W. L. Gibson, Jr., pp. 149–155. Institute for Research on Land and Water Resources: Pennsylvania State University, 1972.

Demsetz, Harold. "The Exchange and Enforcement of Property Rights." *Journal of Law and Economics*, 11 (October 1964): 11–26. Also in *The Economics of Legal Relationships*, ed. by Henry G. Manne, pp. 362–377. St. Paul: West Publishing Co., 1975.

―――. "Some Aspects of Property Rights." *Journal of Law and Economics*, 9 (1966): 61–70. In *The Economics of Legal Relationships*, ed. by Henry G. Manne, pp. 184–193. St. Paul: West Publishing Co., 1975.

―――. *The Market Concentration Doctrine*. Washington: American Enterprise Institute for Public Policy Research, 1973.

Dewhurst, J. Frederic, and associates. *America's Needs and Resources*. New York: The Twentieth Century Fund, 1947.

Eastman, Joyce B. *Economic Institutions to Determine Water Quality*. Master's thesis, Clemson University, 1973.

Ferrar, Terry A., and Whinston, Andrew. "Taxation and Water Pollution Control." *Natural Resources Journal*, 12 (July 1972): 307–329.

Furubotn, Eirik G., and Pejovich, Svetozar. *The Economics of Property Rights*. Cambridge, Mass.: Ballinger Publishing Co., 1974.

Galbraith, John Kenneth. *The Affluent Society*. Boston: Houghton Mifflin Co., 1958.

George, Henry. *The Land Question: What It Involves and How It Alone Can Be Settled*. New York: Doubleday Page and Co., 1904.

George, Henry. *Progress and Poverty*. New York: Robert Schalkenbach Foundation, 1948.

Gillis, Floyd E. *Managerial Economics: Decision Making Under Certainty for Business and Engineering*. Reading, Mass.: Addison-Wesley Publishing Co., 1969.

Gramm, W. Philip, and Maurice, S. Charles. "The Economics of Production Holdbacks for Non-Augmentable Resources: Theory and Evidence." Unpublished manuscript, 1975.

Gras, N. S. B., and Larson, Henrietta M. *Casebook in American Business History*. New York: Appleton-Century-Croft, 1939.

Harberger, Arnold C. *Project Evaluation*. Chicago: Markham Publishing Co., 1972.

Haveman, Robert. "Common Property, Congestion and Environmental Pollution." *Quarterly Journal of Economics*, 87: 2 (May 1973): 278–287.

Head, John G. *Public Goods and Public Welfare*. Durham, N.C.: Duke University Press, 1974.

_____. "Public Goods and Public Welfare." *Rivista di diritto finazziario e scienza della finance*, 28: 3 (September 1969): 384–414.

Heilbroner, Robert. *An Inquiry Into the Human Prospect*. New York: W. W. Norton, 1974.

Hite, James C.; Macaulay, Hugh H.; Stepp, James M., and Yandle, Bruce Jr. *The Economics of Environmental Quality*. Washington: American Enterprise Institute for Public Policy Research, 1972.

Hixson, Stephen Lee. *The South Carolina Oyster Industry: An Economic Analysis of Institutional Arrangements*. Master's thesis, Clemson University, 1975.

Hoffman, Peter M. "Evolving Judicial Standards Under the National Environmental Policy Act and the Challenge of the Alaska Pipeline." *Yale Law Journal*, 81: 1592–1939. In *Environmental Law Review–1973*. New York: Clark Boardman Company, 1973.

Howe, Charles W. *Benefit-Cost Analysis for Water System Planning*. Washington: American Geophysical Union, 1971.

Humphrey, Hubert H. "Guaranteed Jobs for Human Rights." *The Annals*, 418 (March 1975): 17–25.

Jacobellis v. Ohio, 378 U.S. 184, decided June 22, 1964.

Jukes, Thomas H. "DDT, Human Health and the Environment." *Environmental Affairs*, 19 (1971): 535–564. In *Effects of DDT on Man and Other Mammals*. New York: MSS Information Corporation, 1973.

Kamien, M. I.; Schwartz, N. L., and Dolbear, F. T. "Asymmetry Between Bribes and Charges." *Water Resources Research*. 2: 1 (1st Quarter 1966): 147–157.

Kneese, Allen V., *The Economics of Regional Water Quality Manage-*

*ment*. Baltimore: The Johns Hopkins University Press, 1964.

_____. "Research Goals and Progress Toward Them." In *Environmental Quality in a Growing Economy*, ed. by Henry Jarrett, 69–87. Resources for the Future, Baltimore: The Johns Hopkins Press, 1966.

Kneese, Allen V., and Bower, Blair T. *Managing Water Quality: Economics, Technology, Institutions*. Baltimore: Johns Hopkins Press, 1968.

Kneese, Allen V., and Schultze, Charles L. *Pollution, Prices, and Public Policy*. Washington: The Brookings Institution, 1975.

Knight, Frank H. "Some Fallacies in the Interpretation of Social Cost." *Quarterly Journal of Economics*, 38 (1924): 582–606. In Stigler, George J., and Boulding, Kenneth E. *Readings in Price Theory*. Chicago: Richard D. Irwin, 1952.

Kohler, Heinz. *Economics: The Science of Scarcity*. Hinsdale, Ill.: Dryden Press, 1970.

Macaulay, Hugh H. "Environmental Quality, the Market, and Public Finance." In *Modern Fiscal Issues*, ed. by R. M. Bird and J. G. Head, pp. 187–224. Toronto: University of Toronto Press, 1972.

_____. *Economic Effects of Subsidies for Waste Abatement*. Water Resources Research Institute, Clemson University, Report No. 43, May 1974.

_____. "An Evaluation of Subsidies for Water Pollution Abatement." In *The Economics of Federal Subsidy Programs*, pp. 1018–1039. Joint Economic Committee of the U.S. Congress. 1974.

Macaulay, Hugh H., and Yandle, T. Bruce, Jr. *An Economic Evaluation of Water Quality Management Systems*. Water Resources Research Institute, Clemson University, Report No. 58, June 1975.

Maddox, John. *The Doomsday Syndrome*. New York: McGraw-Hill Book Co. 1972.

Manne, Henry G. *The Economics of Legal Relationships*. St. Paul: West Publishing Co. 1975.

Mar, B. W. "A System of Waste Discharge Rights for the Management of Water Quality." *Water Resources Research*, 7: 5 (October 1971): 1079–1086.

Marshall, Alfred. "Three Lectures on Progress and Poverty by Alfred Marshall." *Journal of Law and Economics*, 12: 1 (August 1969): 217–226.

McFarland, William B. "Strategies in Water Quality Control." *Natural Resources Journal*, 12: 3 (July 1972): 318–329.

McKean, Roland N. "Property Rights, Appropriability, and Externalities in Government." In *Perspectives on Property*, ed. by Gene Wunderlich and W. L. Gibson, Jr., pp. 32–55. The Institute for Research on Land and Water Resources: Pennsylvania State University, 1972.

McKetta, John J. "8 Surprises." *Houston Chronicle*, 17 September, 1974, p. 19.

Mills, Edwin S., and Peterson, Frederick M. "Environmental Quality: The First Five Years." *American Economic Review*, 65: 3 (June 1975): 259–268.

Mishan, E. J. "Reflections on Recent Developments in the Concept of External Effects." *Canadian Journal of Economics and Political Science*, 31 (February 1965): 3–34.

―――――. "The Post-War Literature on Externalities: An Interpretative Essay." *Journal of Economic Literature*, 9: 1 (March 1971): 1–28.

Mohring, Herbert, and Boyd, J. H. "Analyzing 'Externalities': 'Direct Interaction' vs. 'Asset Utilization' Framework." *Economica*, 58 (November 1971): 347–361.

Montgomery, W. David. "Markets in Licenses and Efficient Pollution Control Programs." *Social Science Working Paper*, 9 (March 1972): 2–39.

Musgrave, Richard A. *The Theory of Public Finance*. New York: McGraw-Hill Book Co., 1959.

Nelli, H. O. "The Earliest Insurance Contract: A New Discovery." *Journal of Risk and Insurance*, XXXIX: 2 (June 1972): 215–220.

*New York Times*. "1966 Heat Wave Linked to Fatal Strokes." 11 August 1971, p. 15.

*New York Times Index*.

Nordhaus, William, and Tobin, James. *Is Growth Obsolete*: Fiftieth Anniversary Colloquium V, National Bureau of Economic Research, New York: Columbia University Press, 1972.

Nozick, Robert. *Anarchy, State, and Utopia*. New York: Basic Books, 1974.

Okun, Arthur M. *Equality and Efficiency: The Big Trade Off*. Washington, D.C.: The Brookings Institution, 1975.

*Oxford English Dictionary, Compact Edition*. New York: Oxford University Press, 1971.

Peach, Dr. W. N. "The Energy Outlook for the 1980s." U.S. Congress, Joint Economic Committee, 93rd Congress, 1st sess., December 17, 1973.

Pecora, William T. "Science and the Quality of Our Environment." *Bulletin of the Atomic Scientists*, 26: 8 (October 1970): 20–23.

Pejovich, Svetozar. "Towards a General Theory of Property Rights." *Zeitschrift für Nationalökonomie*, 31 (1971): 141–155. In *The Economics of Property Rights*, ed. by Eirik G. Furobotn and Svetozar Pejovich, 341–353, 1974.

Pigou, A. C. *The Economics of Welfare*. London: Macmillan and Co., 1920.

Pirsig, Robert M. *Zen and the Art of Motorcycle Maintenance*. Toronto: Bantam Books, 1974.

Rouse, James W. "Cities That Work for Men—Victory Ahead." An address delivered in San Juan, Puerto Rico, October 25–28, 1971. As reported in Judi Davis, "Some Economic Considerations of Privately Controlled New Towns," unpublished manuscript, Clemson University, 28 April 1973.

Samuels, Warren J. "Welfare Economics, Power, and Property." In *Perspectives on Property*, ed. by Gene Wunderlich and W. L. Gibson, Jr., Institute for Research on Land and Water Resources: Pennsylvania State University, 1972, pp. 61–148.

Samuelson, Paul A., *Economics*. 9th ed. New York: McGraw-Hill Book Co., 1973.

Schmid, A. Allen. "Nonmarket Values and Efficiency of Public Investments in Water Resources." *American Economic Review*, 57: (May 1967): 158–168.

Shepherd, William G. *Economic Performance Under Public Ownership: British Fuel and Power*. New Haven: Yale University Press, 1965.

Siegan, Bernard H. "Non-zoning in Houston." *Journal of Law and Economics*, 13: 1 (April 1970): 71–147.

Smith, Adam. *The Theory of Moral Sentiments*. New York: Augustus M. Kelley, 1966.

Stigler, George J. "Alfred Marshall's Lectures on Progress and Poverty." *Journal of Law and Economics*, 12: 1 (April 1969): 205.

Stigler, George J., and Friedland, Claire. "What Can Regulators Regulate? The Case of Electricity." *Journal of Law and Economics*, 5 (1962): 1–16.

Stone, Christopher D. "Should Trees Have Standing?—Toward Legal Rights for Natural Objects." *The Southern California Law Review*, 45: 2: 450–501. Reprinted in *Environmental Law Review—1973*, pp. 553–604.

Tarlock, A. Dan. "Notes for a Revised Theory of Zoning." In *Perspectives on Population*, ed. by Gene Wunderlich and W. L. Gibson, Jr., Institute for Research on Land and Water Resources: Pennsylvania State University, 1972.

Tideman, T. Nicolaus. "Property as a Moral Concept." In *Perspectives on Property*, ed. by Gene Wunderlich and W. L. Gibson, Jr., The Institute for Research on Land and Water Resources: Pennsylvania State University, 1972.

Tobin, James. "On Limiting the Domain of Inequality." *Journal of Law and Economics*, 13 (October 1970): 263–277.

Townsend, James G. "Investigation of the Smog Incident in Donora, Pa., and Vicinity." *American Journal of Public Health*, 40 (February 1950): 183–189.

Train, Russell E. "Taking Charge of Our Future." Remarks before the Public Affairs Outlook Conference of the Conference Board, New York City, 18 March 1975.

Tripp, James T. B., and Hall, Richard M. "Federal Enforcement Under the Refuse Act of 1899." *Albany Law Review*, 35 (1970): 60. In *Environmental Law Review*, ed. by Floyd Sherrod, pp. 529–553, 1971.

Turvey, Ralph. "On Divergencies Between Social Cost and Private Cost." *Economica*, 30 (August 1963): 309–313.

U.S., Bureau of the Census, *Statistical Abstract of the United States, 1974*.

U.S., Congress, Joint Economic Committee, *The Analysis and Evaluation of Public Expenditures: The PPB System*, 91st Cong. 1st sess., 1969.

U.S., Congress, Joint Economic Committee, *Hearings on Economic Analysis and the Efficiency of Government*, 1971.

U.S., Council on Environmental Quality, *Environmental Quality* 4th Annual Report, September 1973.

U.S., Environmental Protection Agency, *The Economics of Clean Water, 1972*.

U.S., National Water Commission, *Water Policies for the Future*, June 1973.

U.S., Senate, Committee on Public Works, Subcommittee on Air and Water Pollution, *Hearings on Water Pollution, 1970*, 91st Cong. 2d sess., 20 April 1970.

Vickrey, William S. "Optimization of Traffic and Facilities." *Journal of Transport Economics and Policy*, May 1967: 123–136.

Viner, Jacob. "Cost Curves and Supply Curves." *Zeitschrift fur Nationalokonomie*, 3 (1931): 23–46. In Stigler, George J., and Boulding, Kenneth E. *Readings in Price Theory*. Chicago: Richard D. Irwin, 1952.

*Wall Street Journal*. "Companies Complain that Pollution Laws Conflict, Change Often." 23 December 1970, pp. 1, 15.

*Webster's New International Dictionary of the English Language*. Unabridged. Springfield, Mass.: G. & C. Merriam Co. 2nd ed., 1959; 3rd ed. 1976.

Weiss, Shirley E. "New Town Development in the United States: Public Policy and Private Entrepreneurship." South African Institute for Public Administration, 4: 3 (June 1969).

Whyte, William H. *The Last Landscape*. Garden City, N.Y.: Doubleday & Co., 1968.

Wilcox, Clair. *Public Policies Toward Business*. Homewood, Ill.: Richard D. Irwin, 1971.

Worcester, Dean A., Jr. "Pecuniary and Technological Externality, Factor Rents, and Social Costs." *American Economic Review*, 59: 5 (December 1969): 873–885.

Wriston, Walter B. "Whale Oil, Baby Chicks, and Energy." *National Review*, 7 June 1974, pp. 643–646.

Wunderlich, Gene. "Some Basics in Land-Use Policy." In *Land-Use Issues*, ed. by J. Payton Marshall and Peter M. Ashton, pp. 6–12. Blacksburg: Cooperative Extension Service, Virginia Polytechnic Institute and State University, Pub. 69.

Yandle, Bruce, Jr. "Externalities and Highway Location." *Traffic Quarterly*, 24: 4 (October 1970): 583–589.

_____. "Property in Price." *Journal of Economic Issues*, IX: 3 (September 1975): 501–514.

Yandle, Bruce, Jr., and Barnett, A. H. "Henry George, Property Rights, and Environmental Quality." *The American Journal of Economics and Sociology*, 33: 4 (October 1974): 393–400.

Young, J. Z. *Doubt and Certainty in Science*. Oxford: Clarendon Press, 1951.

Zolar. *The History of Astrology*. New York: Arco Publishing Co., 1972.

# Index

# Index

## About the Authors

**Hugh H. Macaulay** received the B. S. degree from the University of Alabama in 1947 and the Ph.D. degree in economics from Columbia University in 1957. Since 1949 he has been a faculty member at Clemson University except for two years (1959-60) when he was a fiscal economist on the tax analysis staff of the Treasury Department. Macaulay has also been a visiting professor of economics at Texas Tech University (1972) and Texas A&M University (1974). His primary research interests and publications have related to the tax treatment of fringe benefits and environmental economics. Macaulay is currently Alumni Professor of Economics and Industrial Management at Clemson University and visiting professor of economics at Texas A&M University (1977-78).

**Bruce Yandle** graduated from Mercer University (Macon, Ga.) in 1955. After spending a number of years in business, he entered Georgia State University where he received the M. B. A. degree (1968) and the Ph.D. degree in economics in 1970. Since 1969 he has been on the faculty of the Department of Economics at Clemson University, serving as head of the department since 1972. Yandle was a senior economist on the staff of the Council on Wage and Price Stability in 1976-77, working primarily on the economic impact of environmental regulations. His research and writing have related primarily to microeconomic topics on housing, environmental resources, and industrial organization. He is currently an associate professor of economics at Clemson University.